T0185400

A Primer on Electromagnetic Fields

Fabrizio Frezza

A Primer on Electromagnetic Fields

 Springer

Fabrizio Frezza
Department of Information Engineering,
    Electronics and Telecommunications
"La Sapienza" University of Rome
Rome
Italy

ISBN 978-3-319-36326-4   ISBN 978-3-319-16574-5   (eBook)
DOI 10.1007/978-3-319-16574-5

Springer Cham Heidelberg New York Dordrecht London
© Springer International Publishing Switzerland 2015
Softcover reprint of the hardcover 1st edition 2015

Printed on acid-free paper

Springer International Publishing AG Switzerland is part of Springer Science+Business Media
(www.springer.com)

In memory of
Prof. Giorgio Gerosa (1931–2013)

Preface

This text collects the material delivered in the course "Electromagnetic Fields" that I lectured starting from the academic year 2003–2004. This course was entitled with a value of five credits for the Science for Engineering Master Degree.

The book was first written with the intention of producing a text of a small size (compared to the reference book by G. Gerosa and P. Lampariello [1]) that would retain most of the important application topics and, at the same time, a rigorous analytic treatment of all the arguments exposed. This last requirement was particularly important as the Master Degree in Science for Engineering aimed at providing all students with a solid and rigorous background in physics and mathematics.

Over time I tried to improve the text usability, keeping the information self-consistent and reporting all the fundamental intermediate steps in analytical computations. I now believe that, because of its conciseness, this book can be a very useful aid for all Electromagnetism students.

My teaching experience makes me think that the objective was achieved, thanks both to an increasing lecturer effort and to a certain diligence required in the exam preparation, as students themselves recognized.

I found it very useful to make the text available on the Internet, for a long time. This approach permitted to implement real-time corrections and additions, it was very convenient for my students. Moreover, the interested reader can take a look at the personal webpage reported in https://web.uniroma1.it/dip_diet/users/frezza, which contains a considerable amount of informative and subsidiary material, as well as details on the topics of this text; everything was personally supervised either by me or by the talented colleagues and co-workers whom I also wish to thank.

The text starts with an introduction to the basic equations and theorems, followed by general and fundamental classic electromagnetic arguments whose value is both practical (for applications) and theoretical: plane waves, transmission lines, waveguides and Green's functions.

I am beholden to my former student, and now Ph.D. in Engineering, Dr. Mauro Mineo, who helped me in composing the text with dedication and expertise, taking care of the successive LATEX versions. I am also deeply indebted to my Ph.D. student, Ing. Patrizio Simeoni, for his great help during the preparation of the English version: without such support, this book could not have been written.

Fabrizio Frezza

Contents

Chapter 1
Fundamental Theorems and Equations of Electromagnetism

Abstract After some elements about differential operators and their analytical properties, the general basic characteristics of electromagnetic fields are described, both in time and in frequency domain. Maxwell's equations, boundary conditions, constitutive relations are treated. The material media properties are investigated and an introductory treatment of dispersion is given. The fundamental Poynting and uniqueness theorems are derived. The wave equation is obtained. Finally, electromagnetic potentials are introduced.

1.1 Properties of the Nabla Vector Linear Operator

$$\nabla = \underline{x}_o \frac{\partial}{\partial x} + \underline{y}_o \frac{\partial}{\partial y} + \underline{z}_o \frac{\partial}{\partial z}.$$

$$\nabla \phi = \underline{x}_o \frac{\partial \phi}{\partial x} + \underline{y}_o \frac{\partial \phi}{\partial y} + \underline{z}_o \frac{\partial \phi}{\partial z}.$$

$$\nabla \cdot \underline{A} = \left(\frac{\partial}{\partial x} \; \frac{\partial}{\partial y} \; \frac{\partial}{\partial z} \right) \begin{pmatrix} A_x \\ A_y \\ A_z \end{pmatrix} = \frac{\partial A_x}{\partial x} + \frac{\partial A_y}{\partial y} + \frac{\partial A_z}{\partial z}.$$

$$\nabla \times \underline{A} = \begin{vmatrix} \underline{x}_0 & \underline{y}_0 & \underline{z}_0 \\ \frac{\partial}{\partial x} & \frac{\partial}{\partial y} & \frac{\partial}{\partial z} \\ A_x & A_y & A_z \end{vmatrix} = \left(\frac{\partial A_z}{\partial y} - \frac{\partial A_y}{\partial z} \right) \underline{x}_0 + \left(\frac{\partial A_x}{\partial z} - \frac{\partial A_z}{\partial x} \right) \underline{y}_0 + \left(\frac{\partial A_y}{\partial x} - \frac{\partial A_x}{\partial y} \right) \underline{z}_0$$

$$\nabla \times \nabla V = 0 \qquad \nabla \cdot \nabla \times \underline{A} = 0.$$

$\nabla^2 = \nabla \cdot \nabla$, that's the reason why the Laplacian is indicated with ∇^2. Note that such definition is also applicable to the vectorial case, exploiting the definition below for the gradient of a vector and the divergence of a dyad.[1]

---

[1] A dyad is defined in Sect. 1.7.

© Springer International Publishing Switzerland 2015
F. Frezza, *A Primer on Electromagnetic Fields*,
DOI 10.1007/978-3-319-16574-5_1

$\nabla^2 \underline{A} = \underline{x}_o \nabla^2 A_x + \underline{y}_o \nabla^2 A_y + \underline{z}_o \nabla^2 A_z$, only in Cartesian coordinates.

$\nabla \times \nabla \times \underline{A} = \nabla \nabla \cdot \underline{A} - \nabla^2 \underline{A} \Longrightarrow \nabla^2 \underline{A} = \nabla \nabla \cdot \underline{A} - \nabla \times \nabla \times \underline{A}$.

$$\nabla \underline{A} = \begin{pmatrix} \frac{\partial}{\partial x} \\ \frac{\partial}{\partial y} \\ \frac{\partial}{\partial z} \end{pmatrix} \begin{pmatrix} A_x & A_y & A_z \end{pmatrix} = \begin{pmatrix} \frac{\partial A_x}{\partial x} & \frac{\partial A_y}{\partial x} & \frac{\partial A_z}{\partial x} \\ \frac{\partial A_x}{\partial y} & \frac{\partial A_y}{\partial y} & \frac{\partial A_z}{\partial y} \\ \frac{\partial A_x}{\partial z} & \frac{\partial A_y}{\partial z} & \frac{\partial A_z}{\partial z} \end{pmatrix}.$$

$\nabla \cdot \underline{\underline{D}} = \left(\nabla \cdot \underline{D}_x \right) \underline{x}_o + \left(\nabla \cdot \underline{D}_y \right) \underline{y}_o + \left(\nabla \cdot \underline{D}_z \right) \underline{z}_o$, being $\underline{D}_x$, $\underline{D}_y$, $\underline{D}_z$ column vectors of the dyad $\underline{\underline{D}}$.

1.2 Solenoidal and Irrotational Fields

Definition A vector field $\underline{V}$ is said *solenoidal* (or *divergence free*) in a region S when its divergence is zero in all points belonging to S:

$$\nabla \cdot \underline{V} = 0$$

Definition A vector field $\underline{V}$ is said *irrotational* in a region S if its curl is zero in S:

$$\nabla \times \underline{V} = 0$$

Definition A *simple linear connected* region is a region of space where *every* simple (i.e. devoid of multiple points) and closed curve contained in it is the edge of (at least) one open surface fully contained in the same region.

Counterexample: both the open space deprived of a straight line and a torus (donut shaped region) aren't simple linear connected regions.

Theorem *A vector field $\underline{E}$ irrotational in a simple linear connected region can always be expressed as a gradient of a scalar field V:*

$$\nabla \times \underline{E} = 0 \Longrightarrow \underline{E} = -\nabla V.$$

Such a field is said conservative.

Definition A region is *simple surface connected* when there are no gaps, i.e. when every closed surface belonging to it contains only points belonging to the region.

Theorem *A vector field $\underline{B}$ solenoidal in a simple surface connected region can always be expressed as a curl of a vector field.*

$$\nabla \cdot \underline{B} = 0 \Longrightarrow \underline{B} = \nabla \times \underline{A}.$$

Definition A surface S is said to be *simply connected* when any simple and closed curve drawn on it is the edge of a surface which completely belongs to S.

1.3 Fundamental Theorems of Vector Analysis

Gradient Theorem

$$\int_V \nabla \phi \, dV = \oint_S \underline{n} \, \phi \, dS.$$

Divergence Theorem

$$\int_V \nabla \cdot \underline{A} \, dV = \underbrace{\oint_S \underline{n} \cdot \underline{A} \, dS}_{\text{flux of vector } \underline{A}}.$$

Curl Theorem

$$\int_V \nabla \times \underline{A} \, dV = \oint_S \underline{n} \times \underline{A} \, dS.$$

Stokes Theorem

$$\underbrace{\int_S \underline{n} \cdot \nabla \times \underline{A} \, dS}_{\text{curl's flux}} = \underbrace{\oint_s \underline{s}_o \cdot \underline{A} \, ds}_{\text{vector's circulation}}.$$

Green's Lemma (first form)

$$\oint_S \phi \nabla \psi \cdot \underline{n} \, dS = \oint_S \phi \frac{\partial \psi}{\partial n} \, dS = \int_V \left(\nabla \phi \cdot \nabla \psi + \phi \nabla^2 \psi \right) \, dV.$$

Green's Lemma (second form)

$$\oint_S (\phi \nabla \psi - \psi \nabla \phi) \cdot \underline{n} \, dS = \oint_S (\phi \frac{\partial \psi}{\partial n} - \psi \frac{\partial \phi}{\partial n}) \, dS = \int_V (\phi \nabla^2 \psi - \psi \nabla^2 \phi) \, dV.$$

1.4 Physical Dimensions and Notation of the Quantities Treated in the text

$\underline{E}(\underline{r},t)$: Electric field (intensity)	$\frac{volt}{meter}$	$\left[\frac{V}{m}\right]$
$\underline{H}(\underline{r},t)$: Magnetic field (intensity)	$\frac{ampere}{meter}$	$\left[\frac{A}{m}\right]$
$\underline{D}(\underline{r},t)$: Electric induction (or electric displacement)	$\frac{coulomb}{meter^2}$	$\left[\frac{C}{m^2}\right]$
$\underline{B}(\underline{r},t)$: Magnetic induction (or magnetic displacement)	$\frac{weber}{meter^2}$	$\left[\frac{Wb}{m^2}\right]$
$\underline{J}(\underline{r},t)$: Electric current density	$\frac{ampere}{meter^2}$	$\left[\frac{A}{m^2}\right]$
$\rho(\underline{r},t)$: Electric charge density	$\frac{coulomb}{meter^3}$	$\left[\frac{C}{m^3}\right]$

where:

$coulomb = ampere \cdot second$ [A · s]

$weber = volt \cdot second$ [V · s]

1.5 Maxwell's Equations and Continuity Conditions

First Maxwell's Equation:

$$\nabla \times \underline{E} = -\frac{\partial \underline{B}}{\partial t}.$$

The above equation represents the differential or local form of the **Faraday-Neumann induction law** (Integral, global or macroscopic form):

$$\underbrace{\oint_s \underline{s}_o \cdot \underline{E}\, ds}_{\text{circulation of } \underline{E}} = -\frac{d}{dt} \underbrace{\int_S \underline{n} \cdot \underline{B}\, dS}_{\text{flux of } \underline{B}}.$$

The differential relation may be derived from the integral form by applying the Stokes theorem:

$$\oint_s \underline{s}_0 \cdot \underline{E}\, ds = \int_S \underline{n} \cdot \nabla \times \underline{E}\, dS = -\int_S \underline{n} \cdot \frac{\partial \underline{B}}{\partial t}\, dS,$$

where the derivative with respect of the time is put inside the integral assuming that the surface S is invariant over time. At this point for the arbitrariness of S and $\underline{n}$ the following equation is obtained:

$$\nabla \times \underline{E} = -\frac{\partial \underline{B}}{\partial t}.$$

Second Maxwell's Equation:

$$\nabla \times \underline{H} = \underline{J} + \frac{\partial \underline{D}}{\partial t}.$$

This is the differential or local form of **Ampère-Maxwell circulation law**:

$$\underbrace{\oint_S \underline{s_0} \cdot \underline{H} \, ds}_{\text{circulation of } \underline{H}} = \underbrace{\int_S \underline{n} \cdot \underline{J} \, dS}_{\text{flux of } \underline{J}} + \underbrace{\frac{d}{dt} \int_S \underline{n} \cdot \underline{D} \, dS}_{\text{flux of } \underline{D}}.$$

The flux of $\underline{J}$, in the above formula, is equal to the conduction or convection current I. The equation can be switched from the integral relation to the differential one in a similar way to what was done for the first Maxwell's equation.

Continuity equation for the electric current:

$$\nabla \cdot \underline{J} = -\frac{\partial \rho}{\partial t}.$$

The formula above is the differential or local form of the **electric charge conservation law**:

$$\underbrace{\oint_S \underline{n} \cdot \underline{J} \, dS}_{\text{electric current}} = \underbrace{-\frac{d}{dt} \int_V \rho \, dV}_{\text{total charge enclosed in V}}.$$

The differential equation can be obtained from the integral relation by applying the divergence theorem and assuming the invariance of the volume V over the time t:

$$\oint_S \underline{n} \cdot \underline{J} \, dS = \int_V \nabla \cdot \underline{J} \, dV = -\int_V \frac{\partial \rho}{\partial t} \, dV,$$

Now, from the arbitrariness of V it follows:

$$\nabla \cdot \underline{J} = -\frac{\partial \rho}{\partial t}.$$

In all the cases shown the local formula was derived from the integral relation, but, performing all the steps backwards, the opposite procedure could also be applied.

The three equations shown up to now are independent. The third and fourth Maxwell's equations can be derived from those, but only in the dynamic case (i.e. when fields are variable in time). In fact, starting from the

$$\nabla \times \underline{E} = -\frac{\partial \underline{B}}{\partial t}$$

and applying the divergence operator to both members the following is obtained:

$$0 = -\nabla \cdot \left(\frac{\partial \underline{B}}{\partial t} \right).$$

At this point, assuming that the functions are of class C^2, the Schwarz theorem, which permits to interchange the order of taking partial derivatives of a function, can be applied obtaining:

$$\frac{\partial}{\partial t}(\nabla \cdot \underline{B}) = 0 \implies \nabla \cdot \underline{B} = constant.$$

It follows that the constant is zero, assuming that $\underline{B}$ does not exist from an infinite time, and therefore

$$\underbrace{\nabla \cdot \underline{B} = 0}_{\text{Third Maxwell's equation}}.$$

The fourth Maxwell's equation can be obtained starting from the second Maxwell's equation, applying the divergence to both sides, using the continuity equation and finally the Schwarz theorem as it is shown below:

$$0 = \nabla \cdot \underline{J} + \nabla \cdot \left(\frac{\partial \underline{D}}{\partial t} \right) = \frac{\partial}{\partial t}(\nabla \cdot \underline{D} - \rho), \implies \underbrace{\nabla \cdot \underline{D} = \rho}_{\text{Fourth Maxwell's equation}}.$$

It should be noticed that ρ represents the density of the *free* charge, while the *polarization* charge isn't taken into account here.

The Gauss theorem can be simply derived from the fourth Maxwell's equation by applying the divergence theorem:

$$\oint_S \underline{n} \cdot \underline{D} \, dS = \int_V \rho \, dV = Q,$$

being Q the electric charge contained in the volume V.

The third and fourth Maxwell's equations are *independent* of the first two in the static case, and therefore in this case all four Maxwell's equations must be imposed.

1.6 Duality Principle and Impressed Sources

The first two Maxwell's equations are not formally symmetrical; in particular there is a missing term in the first equation due to the fact that the magnetic charge, the so-called magnetic monopole, hasn't be discovered yet in nature and, consequently, there isn't any conduction or convection magnetic current. Some fictitious terms are usually introduced so that the wished formal symmetry can be obtained, and, in particular, a density of magnetic charge ρ_m, (measured in $\left[\frac{\text{Wb}}{\text{m}^3}\right]$) and a density of magnetic current $\underline{J}_m$ (measured in $\left[\frac{\text{V}}{\text{m}^2}\right]$) are introduced. The two quantities are related by the following (fictitious) continuity condition of the magnetic current:

$$\nabla \cdot \underline{J}_m = -\frac{\partial \rho_m}{\partial t}.$$

The introduction of ρ_m and $\underline{J}_m$ makes the first and the third Maxwell's equations formally symmetrical to the second and the fourth, respectively:

$$\nabla \times \underline{E} = -\underline{J}_m - \frac{\partial \underline{B}}{\partial t}$$

$$\nabla \cdot \underline{B} = \rho_m$$

Let us now operate the following substitutions in the above equations:

$$\underline{E} \longrightarrow \underline{H} \quad , \quad \underline{H} \longrightarrow -\underline{E},$$

$$\underline{D} \longrightarrow \underline{B} \quad , \quad \underline{B} \longrightarrow -\underline{D},$$

$$\underline{J} \longrightarrow \underline{J}_m \quad , \quad \underline{J}_m \longrightarrow -\underline{J},$$

$$\rho \longrightarrow \rho_m \quad , \quad \rho_m \longrightarrow -\rho,$$

i.e. by replacing every electrical quantity with the corresponding magnetic one and every magnetic quantity with the opposite of the corresponding electrical one, the system of differential equations shown above is transformed into itself. This property allows us to enunciate the so-called **duality principle** in the solutions of the electromagnetic problems: starting from a solution for the electromagnetic field and operating the substitutions above, a solution of another problem (called the dual problem) is obtained. We will often use this principle, for example, to obtain certain results for the magnetic field from analogous results for the electric field.

A final observation is needed on the densities of charge and current shown in the previous equations. In general currents and charges[2] are considered sources of the

---

[2] For example, the current flowing in a transmitting antenna.

electromagnetic field; however, the currents induced by the field on the conductors[3] need to be considered as well. In the first case these quantities are considered known, and therefore they constitute the known term of a non-homogeneous system of differential equations. In the second case, the quantities are, of course, unknown, because they are depending on the unknown electromagnetic field. In certain problems only terms of the second kind are present, in this case the differential problem is said homogeneous. It is clear that a source for the field must be always present[4] but in the homogeneous case it is outside the region in which the solution is considered. It is also true that the source itself is influenced by the field that it emits, but we neglect these influences. At this point we can split the charges and currents in a part which is impressed and one which is field-dependent in the following way:

$$\rho = \rho_i + \rho_c \quad , \quad \rho_m = \rho_{mi} + \rho_{mc},$$
$$\underline{J} = \underline{J}_i + \underline{J}_c \quad , \quad \underline{J}_m = \underline{J}_{mi} + \underline{J}_{mc}.$$

Moreover, since magnetic charges and currents don't exist in nature, we can impose $\rho_{mc} = 0$ and $\underline{J}_{mc} = 0$. Instead, we need to keep both ρ_{mi} and $\underline{J}_{mi}$ because there are cases in which they can be used to represent different forms of excitation; in this case they are called equivalent magnetic impressed charges and currents respectively. For example, we talk about the so-called "magnetic dipole", but it is actually a thin slot on a thin metal plate, illuminated on one side and emitting on the other. Or, as it is done by applying the so-called equivalence theorem (see by analogy the Huygens principle), a field outside a closed surface can be considered as generated by electric and magnetic equivalent currents placed on the surface itself. In conclusion, the first two Maxwell's equations take the form:

$$\nabla \times \underline{E} = -\underline{J}_{mi} - \frac{\partial \underline{B}}{\partial t} \quad , \quad \nabla \times \underline{H} = \underline{J}_i + \underline{J}_c + \frac{\partial \underline{D}}{\partial t}.$$

1.7 Constitutive Relations

The two curl Maxwell's equations constitute a system of two vectorial equations (or 6 scalar equations) containing 5 unknown vectors (or 15 unknown scalars): $\underline{E}$, $\underline{D}$, $\underline{H}$, $\underline{B}$, $\underline{J}_c$ (the reader should recall that $\underline{J}_i$ and $\underline{J}_{mi}$ are known). Nine scalar equations more are needed in order to find a particular solution for the electromagnetic problem. Moreover it needs to be observed that the introduction of the scalar current continuity equation does not help, as it adds the additional unknown variable ρ.

The so-called constitutive relations need to be introduced to find the missing nine scalar equations. Those relations bind together inductions $\underline{D}$, $\underline{B}$ and the current density $\underline{J}_c$ to the fields $\underline{E}$ and $\underline{H}$. Those equations depend on the nature of the medium in which we search our solutions.

---

[3] For example, the current on the receiving antenna.

[4] Even a resonator, which is the typical homogeneous system (the so-called free oscillations, i.e. without forcing) actually has losses and requires excitation.

The easiest case to treat is a vacuum (note that the air can typically be approximated very well by vacuum), for which the following relationships are valid:

$$\underline{D} = \varepsilon_o \underline{E} \qquad \underline{B} = \mu_o \underline{H} \qquad \underline{J}_c \equiv 0,$$

recalling here the values of the magnetic permeability $\mu_o = 4\pi \cdot 10^{-7} \frac{\text{henry}}{\text{meter}} \left[\frac{\text{H}}{\text{m}}\right]$ (the physical dimension is an inductance per unit length), and of the permittivity $\varepsilon_o \cong \frac{10^{-9}}{36\pi} \frac{\text{farad}}{\text{meter}} \left[\frac{\text{F}}{\text{m}}\right]$ (the physical dimensions is a capacity per unit length). The approximation in the above relation is present because it is assumed that the speed of light in vacuum is $c = \frac{1}{\sqrt{\mu_o \varepsilon_o}} \cong 3 \cdot 10^8 \left[\frac{\text{m}}{\text{s}}\right]$, while the actual value is a bit lower than that.

The constitutive relations are more complicated in material media and they need to involve two additional vectors, known as (intensity of) electric polarization $\underline{P}$ and (intensity of) magnetic polarization or magnetization $\underline{M}$ (note the difference in notation compared to some texts of Physics):

$$\underline{D} = \varepsilon_o \underline{E} + \underline{P},$$

$$\underline{B} = \mu_o \underline{H} + \underline{M}.$$

Moreover, $\underline{J}_c$ would be usually non-zero. Finally, apart from the case of special media (such as the so-called chiral media) or of media in motion, $\underline{P}$ and $\underline{J}_c$ depend only on $\underline{E}$ and they don't depend on $\underline{H}$ while $\underline{M}$ depends only on $\underline{H}$ and doesn't depend on $\underline{E}$; it is always assumed here that $\underline{E}$, $\underline{H}$ are the fundamental vectors, (instead of $\underline{E}$, $\underline{B}$, as happens in some texts) so that in a sense they are considered causes, while the other vectors are considered effects.

The general properties of the materials influence the mathematical nature of the constitutive relations. In particular, it is usually considered the hypothesis of linearity, in other words it is assumed the validity of the principle of superposition of the effects. This assumption allows a mathematical matrix-formalism approach.

Another important property is stationarity, or permanency or time invariance: this means that the characteristics of the medium do not vary over time. The third important property is homogeneity or invariance in space: the characteristics of the medium do not depend on the considered point in space.

The fourth fundamental property is isotropy, which essentially represents a medium whose properties are independent of the direction (on the opposite a typical case of non-isotropy or anisotropy, is represented by crystals, in which privileged directions are apparent). This property can also be expressed in a more operational way (i.e., more related to the mathematical form of the constitutive relations that result from the definition) saying that the effect vector is parallel to the cause vector.

The fifth property is dispersivity, which can be either spatial or temporal. A medium is said to be spatially dispersive if at a given point in space, the effect depends on the value of the cause not only at that point, but also in the surrounding area. Similarly, the medium is dispersive in time when the effect at a given time depends on the value of the cause not only at that time, but also in the past instants (the successive instants are excluded because of causality hypothesis). Note that a dispersive medium in space needs to be also dispersive in time because all physical phenomena propagate with a finite speed. The temporal dispersion is generally more significant than the spatial one in many applications.

Finally, a medium is said dissipative for conductivity if the electrical conductivity $\sigma \left(\frac{siemens}{meter} \ \left[\frac{\Omega^{-1}}{m} \right] \right)$ is non-zero.

In the simplest possible medium, i.e. linear, stationary, homogeneous, isotropic and non-dispersive in time nor in space, the following constitutive relations are valid:

$$\underline{P} = \varepsilon_o \chi_e \underline{E},$$

$$\underline{M} = \mu_o \chi_m \underline{H},$$

where the two dimensionless scalars χ_e and χ_m are called electric and magnetic susceptibility, respectively. It follows for the electric induction:

$$\underline{D} = \varepsilon_o(1 + \chi_e)\underline{E} = \varepsilon_o \varepsilon_r \underline{E} = \varepsilon \underline{E},$$

where $\varepsilon = \varepsilon_o \varepsilon_r$ is the dielectric constant of the medium, while $\varepsilon_r = 1 + \chi_e$ is called relative dielectric constant, and for the magnetic induction it is:

$$\underline{B} = \mu_o(1 + \chi_m)\underline{H} = \mu_o \mu_r \underline{H} = \mu \underline{H},$$

where $\mu = \mu_o \mu_r$ is the magnetic permeability of the medium, while $\mu_r = 1 + \chi_m$ is called relative magnetic permeability.

If the medium is dissipative, then we have:

$$\underline{J_c} = \sigma \underline{E} \rightarrow \text{relation which represents Ohm's law in local form.}$$

In the case of a non-homogeneous medium, the only difference with respect to the relationships seen is the dependence on the position r of at least one among the quantities ε, μ and σ. If we assume that the medium is anisotropic, and we refer, for instance, to the relation between $\underline{D}$ and $\underline{E}$, the consequence is that the permittivity becomes a dyad (i.e. a Cartesian tensor of second order, thus having nine components, being the scalar a zero-order tensor, and the vector a first-order tensor: in a three-dimensional space, in fact, a tensor of order n has 3^n components); its mathematical representation is a 3×3 array. The particular instance of diagonal matrices containing all identical values on the main diagonal coincides with the scalar which is on the diagonal.

So we have $\underline{D} = \underline{\varepsilon} \cdot \underline{E}$, where the scalar product between the dyad $\underline{\varepsilon}$ and the vector $\underline{E}$ is the usual product between matrices. In Cartesian coordinates we can write:

$$\begin{aligned}
\underline{\varepsilon} = {}& \varepsilon_{xx} \underline{x}_o \underline{x}_o + \varepsilon_{xy} \underline{x}_o \underline{y}_o + \varepsilon_{xz} \underline{x}_o \underline{z}_o + \\
& + \varepsilon_{yx} \underline{y}_o \underline{x}_o + \varepsilon_{yy} \underline{y}_o \underline{y}_o + \varepsilon_{yz} \underline{y}_o \underline{z}_o + \\
& + \varepsilon_{zx} \underline{z}_o \underline{x}_o + \varepsilon_{zy} \underline{z}_o \underline{y}_o + \varepsilon_{zz} \underline{z}_o \underline{z}_o,
\end{aligned} \quad (1.1)$$

where the juxtaposition of the two unit vectors indicates the dyadic product, which is the matrix product obtained by multiplying a column vector for a row vector; the result of the multiplication is a 3×3 matrix. The constitutive relation in matrix form is:

$$\begin{pmatrix} D_x \\ D_y \\ D_z \end{pmatrix} = \begin{pmatrix} \varepsilon_{xx} & \varepsilon_{xy} & \varepsilon_{xz} \\ \varepsilon_{yx} & \varepsilon_{yy} & \varepsilon_{yz} \\ \varepsilon_{zx} & \varepsilon_{zy} & \varepsilon_{zz} \end{pmatrix} \begin{pmatrix} E_x \\ E_y \\ E_z \end{pmatrix}.$$

Similar considerations apply to the permeability and conductivity. It results in general:

$$\underline{B} = \underline{\underline{\mu}} \cdot \underline{H} \qquad \underline{J_c} = \underline{\underline{\sigma}} \cdot \underline{E}.$$

When the medium is non-homogeneous, those dyads become point functions.

Now let us consider a medium which is linear, stationary or not, homogeneous or not, isotropic or not, non-dispersive in space, but dispersive over time. Then the effect at the generic instant t will depend on the value of the cause in all the previous instants t'. Note that in this specific situation, even in the isotropic case (scalar permittivity), the two vector fields $\underline{D}(\underline{r}, t)$ and $\underline{E}(\underline{r}, t)$ won't be generally parallel because $\underline{E}(\underline{r}, t)$ is not the only cause of $\underline{D}(\underline{r}, t)$. We will have a constitutive relation of the type:

$$\underline{D}(\underline{r}, t) = \int_{-\infty}^{t} \underline{\underline{\varepsilon}}(\underline{r}; t, t') \cdot \underline{E}(\underline{r}, t') \, dt',$$

where the product symbol indicates the usual product between matrices.

In the particular stationary case the tensor $\underline{\underline{\varepsilon}}$ doesn't depend separately on t and t', but it depends only on their difference $t - t'$. The relation turns into a convolution integral that exhibits the convenient property that its Fourier transform (the Fourier transform of $\underline{D}$ in this case) is equal to the product (in this case scalar) of the Fourier transform of $\underline{\underline{\varepsilon}}(\underline{r}, t - t')$, with respect to the variable $t - t'$, and the one of $\underline{E}(\underline{r}, t')$ with respect to the variable t'. Note that the convolution integral definition should be extended between $-\infty$ and $+\infty$, but we can easily comply with this requirement assuming that $\underline{\underline{\varepsilon}} \equiv 0$ when $t' > t$.

In the particular homogeneous case the dependence on $\underline{r}$ disappears, while in the particular isotropic case the tensor turns into a scalar.

The case of spatial dispersion implies further integration of spatial type:

$$\underline{D}(\underline{r}, t) = \int_{-\infty}^{t} \int_V \underline{\underline{\varepsilon}}(\underline{r}, \underline{r}'; t, t') \cdot \underline{E}(\underline{r}', t') \, dV' \, dt'.$$

and this is the most general linear constitutive relation. In the particular case of homogeneous medium, $\underline{\underline{\varepsilon}}$ doesn't depend separately on $\underline{r}$ and $\underline{r}'$, but only on their distance $|\underline{r} - \underline{r}'|$. From this point of view, homogeneity, as it was already mentioned,

can be seen as the counterpart of the spatial stationarity. Similar relationships apply
to $\underline{B}$ and $\underline{J_c}$:

$$\underline{B}(\underline{r}, t) = \int_{-\infty}^{t} \int_{V} \underline{\underline{\mu}}(\underline{r}, \underline{r}'; t, t') \cdot \underline{H}(\underline{r}', t') \, dV' \, dt',$$

$$\underline{J_c}(\underline{r}, t) = \int_{-\infty}^{t} \int_{V} \underline{\underline{\sigma}}(\underline{r}, \underline{r}'; t, t') \cdot \underline{E}(\underline{r}', t') \, dV' \, dt'.$$

1.8 Boundary Conditions

The two curl Maxwell's equations, along with the constitutive relations just seen,
which provide the missing nine scalar equations, can now be solved, but, assigned the
impressed sources, they will generally have a variety of solutions. In order to select
the solution of our particular problem it will be necessary to impose the appropriate
initial and boundary conditions, which will allow us to select, through a uniqueness
theorem, only one solution. The initial conditions involve the electromagnetic field
at a certain instant, in all the points of space. The boundary conditions are related to
the field on particular surfaces at any instant. For example, it is known (and it will
be shown later on) that on the surface of a perfect conductor (i.e. a medium having a
virtually infinite conductivity) the condition of cancellation of the tangential electric
field needs to be imposed at any instant. Another typical boundary condition is the
one at infinity which applies to unlimited regions.

Normally it is preferred to work with the Helmholtz or wave equation, which is
of the second order in one variable, instead of using Maxwell's equations, which are
first-order coupled differential equations. The Helmholtz equation is derived from
Maxwell's equations by imposing the requirement of the homogeneity of the medium.
In general, of course, it would be impossible to deal only with a single homogeneous
medium, but it will be more common to deal with a medium constituted by different
interleaved homogeneous media, that we could define "piecewise homogeneous".
The wave equation would then be resolved in each homogeneous medium and then
the conditions at the interfaces between different media would be imposed. The latter
conditions are the so called *continuity conditions*.

It is well known, from basic physics courses, that the tangential component of $\underline{E}$
and the normal component of $\underline{D}$ are preserved at the interface in the static case. The
same happens for the tangential component of $\underline{H}$ and for the normal component of
$\underline{B}$. In the dynamic case the above relations are still verified, but prudence is required
for $\underline{H}$ in the case in which one of the two media is a perfect conductor, as in this case
surface distributions of charges and currents need to be considered.

Abrupt gaps in the characteristics of the media, which would result in discontinu-
ous functions, don't exist in nature (macroscopically, i.e. where quantum effects can
be neglected), and so we can assume that the properties of the medium vary with
continuity in a thin, but finite, transition region, whose thickness is made tend to

zero by means of a limit procedure; the separation surface is assumed stationary. It
needs to be observed, moreover, that the continuity conditions are not independent;
this is due to the fact that the relations for $\underline{D}$ and $\underline{B}$ are derived from the Maxwell's
divergence equations, while those for $\underline{E}$ and $\underline{H}$ are obtained from the curl equations.
Therefore, in the dynamic case, it is necessary and sufficient to impose the latter
ones and the other two are derived as a result, just as it was done for the Maxwell's
equations.

Let's start from the condition for the field $\underline{H}$, which is derived from

$$\nabla \times \underline{H} = \underline{J} + \frac{\partial \underline{D}}{\partial t},$$

and let us consider a cylindrical region of transition having height $2h$, total area S_t,
unity normal (external) vector $\underline{n}_t$, lateral surface S_3 and volume V, located on either
side of the interface. On the base surface S_1 the characteristics are those of medium
1, on the other base S_2 they are those of medium 2. By integrating the Maxwell's
equation over the volume V and applying the curl theorem, it is obtained:

$$\int_V \nabla \times \underline{H}\, dV = \oint_{S_t} \underline{n}_t \times \underline{H}\, dS = \int_{S_1} \underline{n}_1 \times \underline{H}_1\, dS + \int_{S_2} \underline{n}_2 \times \underline{H}_2\, dS + \int_{S_3} \underline{n}_3 \times \underline{H}_3\, dS =$$

$$= \int_V \underline{J}\, dV + \int_V \frac{\partial \underline{D}}{\partial t}\, dV$$

which, applying the limit $h \longrightarrow 0$ brings to:

$$S_1, S_2 \longrightarrow S \quad , \quad \underline{H}_2 \longrightarrow \underline{H}^+ \quad , \quad \underline{H}_1 \longrightarrow \underline{H}^-,$$

where the positive side is pointed by the tip of the unit normal $\underline{n}$. It can be further
assumed, as is reasonable, that $\underline{H}$ and $\frac{\partial \underline{D}}{\partial t}$ are *limited* in the transition region, this
permits us to cancel both the integral extended to S_3 and the integral of $\frac{\partial \underline{D}}{\partial t}$. The same
assumption can be applied to the integral of $\underline{J}$ only when none of the two media is
perfectly conducting (σ infinite).

Now let us suppose that medium 1, for example, is a perfect conductor. In this case,
approaching the surface, the current density can assume values which are infinitely
high in infinitesimal thickness; it is known that the product of an infinitesimal times
an infinite can give a finite value, therefore it is obtained:

$$\lim_{h \to 0} \int_V \underline{J}\, dV = \lim_{h \to 0} \int_S \int_{-h}^{+h} \underline{J}\, dh\, dS = \int_S (\lim_{h \to 0} \int_{-h}^{+h} \underline{J}\, dh)\, dS = \int_S \underline{J}_S\, dS,$$

having put $\underline{J}_S = \lim_{h \to 0} \int_{-h}^{+h} \underline{J}\, dh$. $\underline{J}_S$ is called surface current density and it is
measured in $\left[\frac{A}{m} \right]$.

From the above it is obtained:

$$\int_S \underline{n} \times (\underline{H}^+ - \underline{H}^-) \, dS = \int_S \underline{J}_S \, dS,$$

and because of the arbitrariness of S it follows

$$\underline{n} \times (\underline{H}^+ - \underline{H}^-) = \underline{J}_S.$$

Finally, by multiplying by $\underline{n}$ both sides of the equation it follows that:

$$[\underline{n} \times (\underline{H}^+ - \underline{H}^-)] \times \underline{n} = \underline{J}_S \times \underline{n}.$$

Reminding that:

$$\underline{A} \times (\underline{B} \times \underline{C}) \neq (\underline{A} \times \underline{B}) \times \underline{C},$$

but when $\underline{A} \equiv \underline{C}$:

$$\underline{C} \times (\underline{B} \times \underline{C}) = (\underline{C} \times \underline{B}) \times \underline{C} = \underline{C} \times \underline{B} \times \underline{C},$$

and so parentheses are not required. In particular, when $\underline{C}$ is a unit vector $\underline{v}_o$ we have:

$$\underline{v}_o \times \underline{B} \times \underline{v}_o = \underline{B}_\perp,$$

where $\underline{B}_\perp$ is the vector component of $\underline{B}$ orthogonal to $\underline{v}_o$.
So

$$\underline{H}_\tau^+ - \underline{H}_\tau^- = \underline{J}_S \times \underline{n},$$

where the subscript τ indicates the tangential component, therefore the above formula indicates that the tangential magnetic field has a discontinuity at the separation surface between the two media, which is equal to the surface current density rotated by $\pi/2$ on the tangent plane.

Let us stress again that the discontinuity occurs only for perfect conductors which can approximate well metals (gold, silver, copper, …) at microwave frequencies (e.g. around 10 GHz), but not at optical frequencies. Apart from perfect conductors, there are good conductors for which the conduction current is prevalent (much greater) on the displacement current, and good dielectrics for which the opposite occurs; in all these cases the tangential component of $\underline{H}$ is continuous.

From the duality principle and from the fact that no induced magnetic surface current exists in nature, it follows that the tangential electric field (in the absence of impressed magnetic surface currents) is always continuous, i.e.:

$$\underline{n} \times (\underline{E}^+ - \underline{E}^-) = 0 \quad \underline{J}_{mS} \equiv 0$$

or

$$\underline{E}_\tau^+ - \underline{E}_\tau^- = 0.$$

The conditions just presented are valid for every kind of medium as constitutive relations are not involved.

Let us now, for example, suppose that medium 1 is a perfect conductor ($\sigma \to \infty$). It is well known that the field $\underline{E}$ must be identically zero inside the perfect conductor, otherwise a volumetric current $\underline{J} \to \infty$ should result from the local Ohm's law $\underline{J} = \sigma \underline{E}$. The above assumption implies $\nabla \times \underline{E} = 0$ so, in the absence of impressed magnetic currents:

$$\frac{\partial \underline{B}}{\partial t} = 0 \Rightarrow \underline{B} = constant \Rightarrow \underline{B} = 0 \Rightarrow \underline{H} = 0, \ \underline{D} = 0.$$

The following results are finally obtained:

- $\underline{E}^- = 0$ and $\underline{n} \times \underline{E}^+ = 0$, in other words it is $\underline{E}_\tau^+ = 0$ (that can be simply written $\underline{n} \times \underline{E} = 0$, $\underline{E}_\tau = 0$),
- $\underline{n} \times \underline{H}^+ = \underline{J}_S$, or $\underline{H}_\tau^+ = \underline{J}_S \times \underline{n}$ (more easily we can write $\underline{n} \times \underline{H} = \underline{J}_S$, $\underline{H}_\tau = \underline{J}_S \times \underline{n}$),

where the electromagnetic field vectors are evaluated on the surface on the side of medium 2 and the unit vector $\underline{n}$ is directed toward medium 2.

Let us now examine the condition for the electric induction $\underline{D}$. This relation isn't independent of the previous ones, because it follows from the equation $\nabla \cdot \underline{D} = \rho$. This time we are going to apply the divergence theorem to the cylinder, so that we have

$$\oint_{S_t} \underline{n}_t \cdot \underline{D} \, dS = \int_V \rho \, dV,$$

i.e. as already seen

$$\int_{S_1} \underline{n}_1 \cdot \underline{D}_1 \, dS + \int_{S_2} \underline{n}_2 \cdot \underline{D}_2 \, dS + \int_{S_3} \underline{n}_3 \cdot \underline{D}_3 \, dS = \int_V \rho \, dV.$$

Now the $\lim_{h \to 0}$ is performed and it is assumed that $\underline{D}$ is limited in the transition region. The same assumption cannot generally apply to ρ, as it wouldn't be true when one of the two medium is supposed perfectly conductor. In particular:

$$\lim_{h \to 0} \int_V \rho \, dV = \lim_{h \to 0} \int_S \int_{-h}^{+h} \rho \, dh \, dS = \int_S (\lim_{h \to 0} \int_{-h}^{+h} \rho \, dh) \, dS = \int_S \rho_S \, dS,$$

where $\rho_S = \lim_{h \to 0} \int_{-h}^{+h} \rho \, dh$ is called the surface charge density and is measured in $\left[\frac{C}{m^2} \right]$.

From steps already presented, it follows:

$$\underline{n} \cdot (\underline{D}^+ - \underline{D}^-) = \rho_S,$$

where, however, it is $\rho_S \equiv 0$ except in the case of the perfect conductor. By duality, in the absence of magnetic charges, we immediately obtain (in the absence of impressed surface magnetic charges):

$$\underline{n} \cdot (\underline{B}^+ - \underline{B}^-) = 0,$$

being $\rho_{mS} \equiv 0$.

1.9 Polarization of Vectors

In the case in which a vector quantity is a sinusoidal function of time, and its three components all have the same angular frequency ω, we will write:

$$\underline{A}(t) = \underline{x}_o A_x(t) + \underline{y}_o A_y(t) + \underline{z}_o A_z(t),$$

$$A_x(t) = |A_x| \cos(\omega t + \varphi_x) = \mathrm{Re}\left[A_x\, e^{j\omega t}\right],$$

$$A_y(t) = |A_y| \cos(\omega t + \varphi_y) = \mathrm{Re}\left[A_y\, e^{j\omega t}\right],$$

$$A_z(t) = |A_z| \cos(\omega t + \varphi_z) = \mathrm{Re}\left[A_z\, e^{j\omega t}\right],$$

having defined the complex scalar quantities (phasors):

$$A_x = |A_x|\, e^{j\varphi_x} = A_{xR} + j A_{xj},$$

$$A_y = |A_y|\, e^{j\varphi_y} = A_{yR} + j A_{yj},$$

$$A_z = |A_z|\, e^{j\varphi_z} = A_{zR} + j A_{zj}.$$

We can introduce now the complex vector (phasor) $\underline{A}$ as the one having the phasors of components as components:

$$\underline{A} = \underline{x}_o A_x + \underline{y}_o A_y + \underline{z}_o A_z = \underline{A}_R + j \underline{A}_j,$$

where

$$\underline{A}_R = \underline{x}_o A_{xR} + \underline{y}_o A_{yR} + \underline{z}_o A_{zR},$$

$$\underline{A}_j = \underline{x}_o A_{xj} + \underline{y}_o A_{yj} + \underline{z}_o A_{zj}.$$

The relationship between the complex vector and the one in the time domain is the customary, since the space unit vectors are real quantities:

$$\underline{A}(t) = \text{Re}\left[\underline{A}\,e^{j\omega t}\right] = \text{Re}\left[\underline{A}_R\,e^{j\omega t} + j\underline{A}_j\,e^{j\omega t}\right] = \underline{A}_R\,\cos(\omega t) - \underline{A}_j\,\sin(\omega t).$$

The vectors $\underline{A}_R$ and $\underline{A}_j$, which do not depend on time, identify a plane in which the tip of $\underline{A}(t)$ describes a geometric locus as time varies. It can be verified that this locus is an ellipse. To show this, let us consider a Cartesian reference system x, y on the the plane $\underline{A}_R$, $\underline{A}_j$ and having its x axis coincident with the direction of $\underline{A}_R$. Let us denote the oriented angle between the direction of $\underline{A}_R$ and the one of $\underline{A}_j$ as θ. The coordinates of the tip of the vector $\underline{A}(t)$ are given by:

$$x(t) = A_R\,\cos(\omega t) - A_j\,\cos\theta\,\sin(\omega t),$$

$$y(t) = -A_j\,\sin\theta\,\sin(\omega t),$$

which represent the parametric equations of the locus sought.

As usual the variable t needs to be removed in order to obtain the Cartesian equation of the locus. Obtaining $\sin(\omega t)$ from the second equation we have:

$$\sin(\omega t) = -\frac{y}{A_j\,\sin\theta},$$

and substituting in the first equation we get:

$$A_R\,\cos(\omega t) = x - \frac{y}{\sin\theta}\cos\theta = x - y\cot\theta.$$

Squaring we get:

$$A_R^2\,\cos^2(\omega t)$$

$$= A_R^2\left[1 - \sin^2(\omega t)\right] = A_R^2\left(1 - \frac{y^2}{A_j^2\,\sin^2\theta}\right) = x^2 - 2xy\cot\theta + y^2\cot^2\theta$$

and ordering the terms

$$x^2 - 2xy\cot\theta + y^2\left(\cot^2\theta + \frac{A_R^2}{A_j^2\,\sin^2\theta}\right) = A_R^2,$$

which is the equation of a conic section, in the form:

$$ax^2 + bxy + cy^2 = d$$

whose discriminant $b^2 - 4ac$ results:

$$4 \cot^2 \theta - 4 \left(\cot^2 \theta + \frac{A_R^2}{A_j^2 \sin^2 \theta} \right) = -\frac{4 A_R^2}{A_j^2 \sin^2 \theta} < 0,$$

and hence it is an ellipse, as one could already guess, given that the coordinates of the tip should always keep limited (assuming that $\underline{A}$ is the representative vector of a physical quantity). It is then said that the vector $\underline{A}(t)$ is elliptically polarized.

There are two important special cases: when $\theta = \pm\pi/2$ (i.e. $\underline{A}_R \perp \underline{A}_j$) and also $A_R = A_j$ it follows that the equation becomes $x^2 + y^2 = A_R^2$ which represents a circumference. In this case $\underline{A}(t)$ is said circularly polarized. In particular, if $\theta = +\pi/2$, the tip of $\underline{A}(t)$ describes the circle in a clockwise direction, while if $\theta = -\pi/2$ $\underline{A}(t)$ describes the circle counter-clockwise. In the literature this kind of polarization is also referred, respectively, as left or right.

The other notable situation is verified when $\underline{A}_R \times \underline{A}_j = 0$, i.e. when one of the following conditions occurs: $\theta = 0$, $\theta = \pi$, $\underline{A}_R = 0$, or $\underline{A}_j = 0$ (in the latter two cases θ is undetermined), in such cases, the tip of $\underline{A}(t)$ describes a straight line segment (where the ellipse degenerates) and it is said that $\underline{A}(t)$ is linearly polarized.

Note that by multiplying the complex vector (phasor) $\underline{A}$ by a scalar quantity in general complex C, we obtain the phasor of a vector function of time which has the same polarization characteristics of $\underline{A}(t)$. In fact, assuming $\underline{B} = C\underline{A}$, we obtain:

$$\underline{B}(t) = \mathrm{Re}\left[\underline{B}\, e^{j\omega t} \right] = \mathrm{Re}\left[|C|\, e^{j\varphi_c} \underline{A}\, e^{j\omega t} \right] = \mathrm{Re}\left[|C|\, \underline{A}\, e^{j(\omega t + \varphi_c)} \right] =$$

$$= |C|\, \mathrm{Re}\left[\underline{A}\, e^{j\omega(t + \varphi_c/\omega)} \right] = |C|\, \underline{A}\left(t + \frac{\varphi_c}{\omega} \right).$$

Therefore $\underline{B}(t)$ is obtained from $\underline{A}(t)$ by multiplying it by a positive real factor $|C|$ and translating it over time of the amount φ_c/ω. These operations do not change, obviously, the polarization characteristics of the vector. So it is always the vector part (and not the scalar) of a certain quantity the responsible for the polarization.

It is worth noting that our considerations apply to vector fields which are functions of space coordinates and time, so the phasor $\underline{A}$ and the vectors $\underline{A}_R$ and $\underline{A}_j$ are in general functions of the point, e.g., there will be a locus in which the polarization is circular and one in which it will be linear. In the important case of plane waves, however, the polarization is the same in all points of space as will be shown in the next chapter.

In the case of damped oscillatory phenomena (presence of loss mechanisms), which can be represented by a complex angular frequency $\omega = \omega_R + j\omega_j$ the tip of $\underline{A}(t)$ describes a spiral in the plane defined by $\underline{A}_R, \underline{A}_j$. As for a more general time dependence, the locus described is not even a plane curve.

1.10 Maxwell's Equations and Constitutive Relations in Frequency Domain

Let us consider now a more generic, with respect to the sinusoidal one, time dependence of our fields, i.e. let us require the only constraint of being Fourier-transformable, so that we can consider the Fourier transformed fields in place of the time-dependent fields, and therefore we can use the multiplication by the factor $j\omega$ instead of the time-domain derivative; this approach permits to turn differential equations into algebraic ones or into differential equations on other variables. With reference to the electric field, for example, we define the vector Fourier transform in the following way:

$$\underline{E}(\underline{r}, \omega) = \Im\left[\underline{E}(\underline{r}, t)\right] = \int_{-\infty}^{\infty} \underline{E}(\underline{r}, t)\, e^{-j\omega t}\, dt\,;$$

its inverse is:

$$\underline{E}(\underline{r}, t) = \frac{1}{2\pi} \int_{-\infty}^{\infty} \underline{E}(\underline{r}, \omega)\, e^{j\omega t}\, d\omega,$$

where the latter is actually an improper integral form (principal Cauchy integral).

With those transformations, Maxwell's equations become:

$$\begin{cases} \nabla \times \underline{E} = -\underline{J}_{mi} - j\omega\, \underline{B} \\ \nabla \times \underline{H} = \underline{J}_i + \underline{J}_c + j\omega\, \underline{D} \end{cases}.$$

It should be noted that the same form for the equations is obtained in the particular monochromatic case (phasors method); the quantities, however, are conceptually different as they have, for example, different physical dimensions: in fact, while the phasor of $\underline{E}$ has the same dimensions of the vector in time, the Fourier transform of $\underline{E}$ is measured in $[(V/m) \cdot s]$, and so on for other quantities.

Now let's see how the constitutive relations change in the transformed domain. In the case of a medium non-dispersive in time, but in general dissipative, non-homogeneous and anisotropic, we have the relations:

$$\underline{D}(\underline{r}, \omega) = \underline{\underline{\varepsilon}}(\underline{r}) \cdot \underline{E}(\underline{r}, \omega),$$

$$\underline{B}(\underline{r}, \omega) = \underline{\underline{\mu}}(\underline{r}) \cdot \underline{H}(\underline{r}, \omega),$$

$$\underline{J}_c(\underline{r}, \omega) = \underline{\underline{\sigma}} \cdot \underline{E}(\underline{r}, \omega),$$

where the dyads $\underline{\underline{\varepsilon}}$, $\underline{\underline{\mu}}$ and $\underline{\underline{\sigma}}$ are real quantities and independent of ω. It is also verified that $\underline{\underline{\varepsilon}}$ and $\underline{\underline{\mu}}$ are symmetric tensors, and it can be demonstrated that the same property is valid for $\underline{\underline{\sigma}}$ due to symmetry properties of the crystalline media. In the particular case of homogeneity the dependence on $\underline{r}$ disappears, while tensors disappear in the isotropic case.

Let us now consider the case of a medium dispersive in time, it occurs that constitutive relations are much simpler in the transformed domain than in the time domain when the stationarity hypothesis is assumed. In fact, considering for example

$$\underline{D}(\underline{r}, t) = \int_{-\infty}^{\infty} \underline{\underline{\varepsilon}}(\underline{r}, t - t') \cdot \underline{E}(\underline{r}, t') \, dt',$$

which represents a convolution integral (stationary medium), and moving to the transformed domain, the above relation turns into a product of the transformed quantities:

$$\underline{D}(\underline{r}, \omega) = \underline{\underline{\varepsilon}}(\underline{r}, \omega) \cdot \underline{E}(\underline{r}, \omega),$$

where it can be observed that $\underline{\underline{\varepsilon}}(\underline{r}, \omega)$ is a *complex* dyadic function of the point $\underline{r}$ and of *the angular frequency* ω, whose components are the Fourier transforms of the components $\varepsilon_{ij}(\underline{r}, t - t')$ of the tensor $\underline{\underline{\varepsilon}}(\underline{r}, t - t')$ with respect to the variable $t - t'$. The simplifications shown in the homogeneous and isotropic cases still occur but with the essential difference that the dielectric constant is now complex and dependent on ω. Of course, analogous considerations apply to the relations between $\underline{B}$ and $\underline{H}$ and between $\underline{J}_c$ and $\underline{E}$.

In the case of damped phenomena, the angular frequency becomes complex (as we could easily guess) and this doesn't permit the use of the Fourier transform (it is known that the Fourier transform requires the transformed variable to be real). However, in this case it is always possible a mathematical treatment with the use of the Laplace transform, in which the transformed variable is complex.

1.11 Dispersive Media

The constitutive relations seen so far represent the result of the study on the cause-effect relationship that exists between the physical entities. It is shown now a typical example in which the direct examination of the phenomenon allows us to determine the expression of the parameters and in detail of the dielectric constant with reference, therefore, to the relationship existing between $\underline{D}$ and $\underline{E}$.

Let us consider, in particular, the so-called Lorentz model used in the study of dielectric polarization. This model examines a non-polar dielectric medium in which the centers of gravity of the bound electric charges (free charges generate the phenomena of conductivity), both positive and negative, coincide in the absence of applied electric field. Under the action of an electric field $\underline{E} = E\underline{e}_o$ (where $\underline{e}_o$ is the unit vector of the electric field itself) the two centers of gravity move each from the other up to a distance ℓ. This displacement generates a dipole moment $q\ell\underline{e}_o$.

Assuming that there are N identical dipoles per unit volume, the vector intensity of polarization $\underline{P}$, which is the dipole moment per unit volume, is given by $\underline{P} = Nq\ell\underline{e}_o = P\underline{e}_o$. In order to derive the relation between $\underline{P}$ and $\underline{E}$, we need to consider the movement of the charge $-q$ with respect to the charge $+q$. Such a movement takes place in the opposite direction of the applied field (so the displacement of the charge is $\underline{\ell} = -\ell\,\underline{e}_o$) under the action of various forces, and in particular the Coulomb force $-q\,\underline{E} = -q\,E\,\underline{e}_o$, a restoring force that tends to take the two charges back in the equilibrium position, in other words, it tries to make them coincide: this force is, for small values of ℓ, an increasing function of ℓ which can be approximated (Taylor series truncated at the first order, then linearisation) with $-k\,\underline{\ell} = k\,\ell\,\underline{e}_o$, i.e. as an elastic force. Finally it needs to be considered in general also a damping force due to collisions (and thus a dissipative term, which indicates a transfer of energy from one form to another). This damping force can be considered proportional to the velocity $\frac{d\ell}{dt}$ of the charge, and expressed as $-\beta\,\underline{v} = -\beta\,\frac{d\underline{\ell}}{dt} = \beta\,\frac{d\ell}{dt}\underline{e}_o$ (viscous friction).

At this point, by applying the second law of dynamics $\underline{F} = m\,\underline{a}$ (where $\underline{a} = \frac{d^2\underline{\ell}}{dt^2} = -\frac{d^2\ell}{dt^2}\underline{e}_o$) projected in the direction of $\underline{e}_o$ we obtain:

$$-m\frac{d^2\ell}{dt^2} = -q\,E + k\,\ell + \beta\,\frac{d\ell}{dt}.$$

Multiplying by $-N\,q$ and rearranging in terms of the function ℓ and its derivatives, we have:

$$m\,N\,q\frac{d^2\ell}{dt^2} + \beta\,N\,q\frac{d\ell}{dt} + k\,N\,q\,\ell = N\,q^2\,E.$$

At this point we need to remember that $P = N\,q\,\ell$, so it follows:

$$m\,\frac{d^2P}{dt^2} + \beta\,\frac{dP}{dt} + k\,P = N\,q^2\,E,$$

which is a constitutive relation $P = P(E)$ written in the form of a differential equation. This means that if one wants to get the function $P = P(E)$ directly in the time domain one must solve this differential equation, and of course one has to somehow integrate. That explains why the constitutive relations seen before are expressed in integral form.

Here, however, we decide to move to the frequency domain so that our differential equation becomes an algebraic equation

$$(-m\,\omega^2 + j\beta\,\omega + k)P = N\,q^2\,E,$$

being now P and E the transformed quantities. The solution now is straightforward:

$$P = \frac{N\,q^2}{k - m\,\omega^2 + j\beta\,\omega}E = \frac{N\,q^2}{m}\frac{1}{\frac{k}{m} - \omega^2 + j\frac{\beta}{m}\omega}E = \frac{N\,q^2}{m}\frac{1}{(\omega_o^2 - \omega^2) + 2j\alpha\omega}E,$$

having put

$$\omega_o = \sqrt{\frac{k}{m}}$$

called resonance angular frequency (remember the theory of the harmonic oscillator) and

$$2\alpha = \frac{\beta}{m}$$

damping term due to the presence of dissipation (loss phenomena). It is typically $\alpha < \omega_o$.

Introducing the expression of P just found in the transformed constitutive relationship $\underline{D} = \varepsilon_o \underline{E} + \underline{P}$ we get:

$$\underline{D} = \left[\varepsilon_o + \frac{N q^2}{m} \frac{1}{(\omega_o^2 - \omega^2) + 2 j \alpha \omega} \right] \underline{E}.$$

So, in the model considered, the medium is a dielectric which is dispersive in time but stationary, whose dielectric constant is complex and dependent on ω:

$$\varepsilon(\omega) = \varepsilon_o + \frac{N q^2}{m} \frac{(\omega_o + \omega)(\omega_o - \omega) - 2 j \alpha \omega}{(\omega_o + \omega)^2 (\omega_o - \omega)^2 + 4 \alpha^2 \omega^2} = \varepsilon_R(\omega) + j \varepsilon_j(\omega),$$

with

$$\varepsilon_R(\omega) = \varepsilon_o + \frac{N q^2}{m} \frac{(\omega_o + \omega)(\omega_o - \omega)}{(\omega_o + \omega)^2 (\omega_o - \omega)^2 + 4 \alpha^2 \omega^2},$$

$$\varepsilon_j(\omega) = -\frac{N q^2}{m} \frac{2 \alpha \omega}{(\omega_o + \omega)^2 (\omega_o - \omega)^2 + 4 \alpha^2 \omega^2}.$$

After inverse transform of $\varepsilon(\omega)$ we could get the $\varepsilon(t)$ to be used in

$$\underline{D}(\underline{r}, t) = \int_{-\infty}^{t} \varepsilon(t - t') \underline{E}(\underline{r}, t') \, dt'.$$

We observe now that in the angular frequency range $\omega \ll \omega_o$ we have:

$$\varepsilon_R(\omega) \simeq \varepsilon_o + \frac{N q^2}{m} \frac{1}{\omega_o^2} \qquad \varepsilon_j(\omega) \simeq 0,$$

while for $\omega \gg \omega_o$ we have

$$\varepsilon_R(\omega) = \varepsilon_o - \frac{N q^2}{m} \frac{1}{\omega^2} \qquad \varepsilon_j(\omega) \simeq 0,$$

and in particular $\lim_{\omega \to \infty} \varepsilon(\omega) = \varepsilon_o$.

In the proximity of the resonance angular frequency $\omega \simeq \omega_o$ it results:

$$\varepsilon_R(\omega) \simeq \varepsilon_o + \frac{N q^2}{m} \frac{1}{2\omega_o} \frac{\omega_o - \omega}{(\omega_o - \omega)^2 + \alpha^2}$$

Fig. 1.1 $\frac{\varepsilon_R}{\varepsilon_o}$ in the Lorentz model

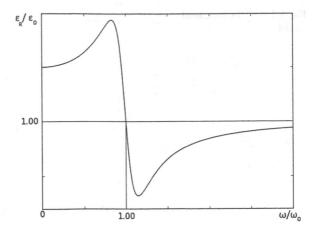

$$\varepsilon_R(\omega_o) = \varepsilon_o,$$

$$\varepsilon_j(\omega) \simeq -\frac{N\,q^2}{m}\frac{1}{2\omega_o}\frac{\alpha}{(\omega_o - \omega)^2 + \alpha^2}$$

$$\varepsilon_j(\omega_o) = -\frac{N\,q^2}{m\,2\omega_o\,\alpha}.$$

The curve of $\frac{\varepsilon_R}{\varepsilon_o}$ as a function of $\frac{\omega}{\omega_o}$ (we should always refer to normalized dimensionless quantities, so that we can avoid dependence on units of measurement and consider pure numbers that have some meaning in themselves) is represented in Fig. 1.1.

ε_R and ε_j are often named in the literature ε' and $-\varepsilon''$; usually ε is expressed as $\varepsilon(\omega) = \varepsilon'(\omega) - j\varepsilon''(\omega)$ to highlight the presence of a negative imaginary part.

Regarding the imaginary part, changed in sign for simplicity in the picture shown below (Fig. 1.2) it is a bell curve centered on ω_o under the hypotheses made, which is called Lorentzian and it is characterized by the fact that it becomes higher and thinner when α (i.e. loss) decreases. This fact recalls us immediately the Dirac function, and in fact we can show that $\lim_{\alpha \to 0}[-\varepsilon_j(\omega)] = \pi\frac{Nq^2}{m2\omega_o}\delta(\omega - \omega_o)$. By the way, it needs to be observed that the previous assumption is true only as a limit procedure, because if $\alpha \equiv 0$, it needs to be $\varepsilon_j(\omega) \equiv 0$.

It is worth noting a general feature: the presence of a non-zero imaginary part is related to loss phenomena different from the ones due to conduction. As a typical example, we can note that this kind of losses are exploited in the microwave-oven heating of substances (for example, food). Moreover, the negative sign of the imaginary part means that there is a power loss, as it will become clear in the subsequent discussion on the complex Poynting's theorem. However, it could be shown that it can never happen that $\varepsilon_j(\omega) \equiv 0$ (which is physically plausible because media with no losses can never exist) due to the so-called Kramers-Krönig relationships, which

Fig. 1.2 $-\frac{\varepsilon_j}{\varepsilon_o}$ in the Lorentz model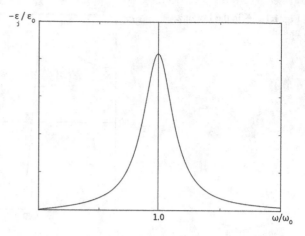

link the real and imaginary part of ε. Those relations state that if the imaginary part were identically zero, then also the real part should be zero. These properties have a very general validity, involving all frequency responses of linear and stationary systems.

Finally, note that the behavior observed for $\omega \to \infty$ ($\varepsilon_R \to \varepsilon_o$ and $\varepsilon_j \to 0$, so $\varepsilon \to \varepsilon_o$) corresponds to the fact that at very high frequencies the elementary dipoles are no longer able to follow the oscillations of the field and the behavior of the dielectric medium tends to the one of a vacuum.

Regarding the magnetic permeability and the constitutive relation that links $\underline{M}$ to $\underline{H}$ one can study for example a medium, said magnetized ferrite, which is anisotropic and dispersive in time, i.e. the elements of the matrix $\underline{\underline{\mu}} = \mu_o \underline{\underline{\mu}}_r$ or at least some of them are in general complex functions of ω. It is in Cartesian coordinates:

$$\underline{\underline{\mu}}_r(\omega) = \begin{pmatrix} \mu_1(\omega) & j\mu_2(\omega) & 0 \\ -j\mu_2(\omega) & \mu_1(\omega) & 0 \\ 0 & 0 & 1 \end{pmatrix},$$

with μ_1, μ_2 complex.

In the particular case of absence of losses the functions $\mu_1(\omega)$ and $\mu_2(\omega)$ become real, and then the matrix $\underline{\underline{\mu}}_r$ becomes a so-called Hermitian matrix, i.e. the elements symmetrical with respect to the main diagonal are complex conjugate (and thus the diagonal elements must be real):

$$a_{ij} = a_{ji}^*$$

Note that in the particular case of real matrices (i.e. the elements of the matrix are real) symmetric and Hermitian matrix coincide. We will encounter again, many times, the association between the hermitianity of the matrix and the absence of losses (this is a general relation).

Just few words on the boundary conditions: they coincide with the ones in the time domain, the only difference is that they are now applied to transformed quantities.

1.12 Poynting's Theorem

We will obtain now an integral relation that needs to be satisfied by any solution of Maxwell's equations: the Poynting's theorem whose energetic interpretation makes it a fundamental of electromagnetism. This theorem can be expressed in both time and frequency domains.

1.12.1 Poynting's Theorem in the Time Domain

Let us consider the time domain first. We start again from the Maxwell's equations:

$$\begin{cases} \nabla \times \underline{E} = -\underline{J}_{mi} - \frac{\partial \underline{B}}{\partial t} \\ \nabla \times \underline{H} = \underline{J}_i + \underline{J}_c + \frac{\partial \underline{D}}{\partial t} \end{cases}$$

Let's consider the scalar product of the first equation by $\underline{H}$, the scalar product of the second equation by $\underline{E}$ and let us subtract the two members. We obtain:

$$\underline{H} \cdot \nabla \times \underline{E} - \underline{E} \cdot \nabla \times \underline{H} = -\underline{J}_{mi} \cdot \underline{H} - \underline{J}_i \cdot \underline{E} - \underline{J}_c \cdot \underline{E} - \underline{H} \cdot \frac{\partial \underline{B}}{\partial t} - \underline{E} \cdot \frac{\partial \underline{D}}{\partial t}.$$

Let us recall now the vectorial relation for the divergence of a vector product $\nabla \cdot (\underline{A} \times \underline{B}) = \underline{B} \cdot \nabla \times \underline{A} - \underline{A} \cdot \nabla \times \underline{B}$. The first member of the equality just seen is then $\nabla \cdot (\underline{E} \times \underline{H})$. We can now integrate in a generic volume V, bounded by a closed surface S, and then we can apply the divergence theorem to the first member of the equation obtaining:

$$\oint_S \underline{n} \cdot \underline{E} \times \underline{H} \, dS + \int_V \underline{J}_c \cdot \underline{E} \, dV + \int_V \left(\underline{E} \cdot \frac{\partial \underline{D}}{\partial t} + \underline{H} \cdot \frac{\partial \underline{B}}{\partial t} \right) dV =$$

$$= -\int_V \left(\underline{J}_i \cdot \underline{E} + \underline{J}_{mi} \cdot \underline{H} \right) dV. \tag{1.2}$$

The quantity $\underline{E} \times \underline{H} = \underline{P}$ has the physical dimensions of $\left[\frac{V \cdot A}{m^2} \right] = \frac{watt}{meter^2} \left[\frac{W}{m^2} \right]$, and hence of a power per unit area, i.e. a surface power density: this is the so-called Poynting vector.

The quantity $\underline{J}_c \cdot \underline{E}$ physically represents the power per unit volume (power density, $\left[\frac{W}{m^3} \right]$) provided by the electromagnetic field to the conduction current density $\underline{J}_c$. This power is dissipated by Joule effect, and then converted into heat. In fact, let us recall here the expression of the Lorentz force acting on a point charge q, which moves with velocity $\underline{v}$:

$$\underline{F} = q \left(\underline{E} + \underline{v} \times \underline{B} \right).$$

The power delivered to the charge by the electromagnetic field is given by the product

$$P = \underline{F} \cdot \underline{v} = q\,(\underline{v} \cdot \underline{E} + \underline{v} \cdot \underline{v} \times \underline{B}) = q\,\underline{v} \cdot \underline{E},$$

having applied the mixed product property: $\underline{a} \cdot \underline{b} \times \underline{c} = \underline{a} \times \underline{b} \cdot \underline{c}$. In a similar way, the charge density ρ will be subject to a force density $\underline{f}$ (force per unit volume) given by $\rho(\underline{E} + \underline{v} \times \underline{B})$, which corresponds to the power density provided by the electromagnetic field: $p_c = \rho \underline{v} \cdot \underline{E} = \underline{J}_c \cdot \underline{E}$ being $\underline{J}_c = \rho \underline{v}$ the current density (convection) represented by the charge density in motion.

Similarly, the term $p_i = -\underline{J}_i \cdot \underline{E}$ expresses, for the action-reaction-law (by virtue of the minus sign), the power density supplied to the field from the impressed electric current density. And for duality the term $p_{mi} = -\underline{J}_{mi} \cdot \underline{H}$ represents the power density supplied to the field from the impressed magnetic current density.

In the case of an isotropic medium, non-dispersive, but dissipative for conductivity, we have:

$$p_c = \sigma\,\underline{E} \cdot \underline{E} = \sigma\,E^2 = \sigma(E_x^2 + E_y^2 + E_z^2).$$

So p_c is a positive definite quadratic form in the components of $\underline{E}$. Indeed, for a passive medium (i.e. which does not simulate a generator, for which $\sigma > 0$) we have $p_c > 0$ and $p_c = 0 \Leftrightarrow E_x = E_y = E_z = 0$.

In the case of an anisotropic medium the conductivity is a tensor and so we have:

$$p_c = (\underline{\underline{\sigma}} \cdot \underline{E}) \cdot \underline{E} = \sum_{i=1}^{3}\left(\sum_{j=1}^{3} \sigma_{ij}\,E_j\right) E_i = \sum_{i=1}^{3}\sum_{j=1}^{3} \sigma_{ij}\,E_i\,E_j,$$

p_c is again a quadratic form in the components of $\underline{E}$, which must necessarily be positive definite for a passive medium. This is equivalent, because of the properties of quadratic forms, to require that all the eigenvalues of the matrix $\underline{\underline{\sigma}}$ are positive (and therefore first of all real) in passive media.

Let's now consider the term $p_E = \underline{E} \cdot \frac{\partial \underline{D}}{\partial t}$. For non-dispersive and isotropic media we have:

$$p_E = \varepsilon\,\underline{E} \cdot \frac{\partial \underline{E}}{\partial t} = \frac{\partial}{\partial t}\left(\frac{1}{2}\,\varepsilon\,\underline{E} \cdot \underline{E}\right) = \frac{\partial}{\partial t}\left(\frac{1}{2}\,\underline{E} \cdot \underline{D}\right) = \frac{\partial}{\partial t}\left(\frac{1}{2}\,\varepsilon\,E^2\right).$$

The quantity $w_E = \frac{1}{2}\,\underline{E} \cdot \underline{D}$ represents, as it is well known from the basic physics courses, the density of electrical energy $\left(\frac{joule}{meter^3}\right)\left[\frac{J}{m^3}\right]$ stored, i.e., the electrical energy stored in a unit volume. Note that w_E depends only on the present value of $\underline{E}$ (so it is a state function) and it is a positive definite quadratic form in the components of $\underline{E}$. It follows:

$$p_E = \frac{\partial w_E}{\partial t},$$

and therefore p_E represents the density of stored electrical power.

From the duality relations, the term

$$p_H = \underline{H} \cdot \frac{\partial \underline{B}}{\partial t}$$

is the stored magnetic power density. For non-dispersive and isotropic media we have:

$$p_H = \mu \underline{H} \cdot \frac{\partial \underline{H}}{\partial t} = \frac{\partial}{\partial t} \left(\frac{1}{2} \mu \underline{H} \cdot \underline{H} \right) = \frac{\partial}{\partial t} \left(\frac{1}{2} \underline{H} \cdot \underline{B} \right) = \frac{\partial}{\partial t} \left(\frac{1}{2} \mu H^2 \right).$$

The quantity $w_H = \frac{1}{2} \underline{H} \cdot \underline{B}$ represents the stored magnetic energy density. It depends only on the present value of $\underline{H}$ (it is again a state function) and it is a positive definite quadratic form in the components of $\underline{H}$. We have therefore:

$$p_H = \frac{\partial w_H}{\partial t}.$$

All relations seen continue to be valid in the anisotropic case, but we need to insert the relevant tensors:

$$w_E = \frac{1}{2} \underline{E} \cdot \left(\underline{\underline{\varepsilon}} \cdot \underline{E} \right) = \frac{1}{2} \sum_{i=1}^{3} E_i \left(\sum_{j=1}^{3} \varepsilon_{ij} E_j \right) = \frac{1}{2} \sum_{i=1}^{3} \sum_{j=1}^{3} \varepsilon_{ij} E_i E_j,$$

and similarly:

$$w_H = \frac{1}{2} \underline{H} \cdot \left(\underline{\underline{\mu}} \cdot \underline{H} \right) = \frac{1}{2} \sum_{i=1}^{3} \sum_{j=1}^{3} \mu_{ij} H_i H_j.$$

These quadratic forms must still be positive definite, due to their physical meaning of energy, and this assumption puts constraints on the tensors $\underline{\underline{\varepsilon}}$, $\underline{\underline{\mu}}$, since it is required that their eigenvalues must all be positive.

In addition, it can be shown that these tensors must be symmetric (i.e. symmetric real matrices, and therefore real eigenvalues), so that the energies become in fact state functions (dependence only on the initial and final states, but not on the transformation applied, or, the absence of hysteresis phenomena).

In dispersive media, finally, it is not possible to define energy as a state function.

It remains to interpret the meaning of the term that expresses the flux of the Poynting vector through a closed surface S. Given the interpretation of the other terms, actually, the Poynting's theorem expresses the power balance of the electromagnetic field. The right member of the equality represents the power provided by impressed currents (i.e. by the sources) to the electromagnetic field. For the principle of energy conservation the various terms of the left member express the uses or destinations of such power. In particular the flux of $\underline{P}$ represents the outgoing (or incoming) power through the closed surface S.

Note that, however, it is wrong in general to attribute to the quantity $\underline{n} \cdot \underline{P}$ the meaning of power that passes through the unit area perpendicular to the direction of $\underline{n}$. In order to show this, it is sufficient to consider a static field generated by permanent magnets and electrostatic charges (hence no current). In this case fields $\underline{E}$, $\underline{H}$ will generally be non-zero and non-parallel. Therefore, the flux of $\underline{P}$ through an open unit surface will be generally different from zero. This does not, however, correspond to a radiated power, as the origin of the field is static. The flux of $\underline{P}$ through a closed surface is in fact zero. We have:

$$\nabla \times \underline{E} = 0 \quad \nabla \times \underline{H} = 0 \ \Rightarrow \nabla \cdot \underline{P} = 0 \ \Rightarrow \oint_S \underline{n} \cdot \underline{P} \, dS = 0.$$

Finally note that:

$$\underline{n} \cdot \underline{P} = \underline{n} \cdot \underline{E} \times \underline{H} = \underline{n} \cdot \underline{E}_T \times \underline{H}_T,$$

so there can be no power flux inside a perfect conductor ($\underline{E}_T = 0$). The role of conductors in electrical power lines is, as known, only to guide the energy, which is actually transmitted through the interposed dielectric.

1.12.2 Poynting's Theorem in the Frequency Domain

Let us now consider the Poynting's theorem in thefrequency domain. We start this time from the transformed Maxwell's equations:

$$\begin{cases} \nabla \times \underline{E} = -\underline{J}_{mi} - j\omega \, \underline{B} \\ \nabla \times \underline{H} = \underline{J}_i + \underline{J}_c + j\omega \, \underline{D} \end{cases}.$$

We scalar multiply the first equation by $\underline{H}^*$, while we conjugate the second[5] and then multiply it by $\underline{E}$:

$$\underline{H}^* \cdot \nabla \times \underline{E} = -\underline{H}^* \cdot \underline{J}_{mi} - j\omega \, \underline{H}^* \cdot \underline{B},$$

$$\underline{E} \cdot \nabla \times \underline{H}^* = \underline{E} \cdot \underline{J}_i^* + \underline{E} \cdot \underline{J}_c^* - j\omega \, \underline{E} \cdot \underline{D}^*.$$

Subtracting member to member, we have:

$$\underline{H}^* \cdot \nabla \times \underline{E} - \underline{E} \cdot \nabla \times \underline{H}^* = -\underline{J}_{mi} \cdot \underline{H}^* - \underline{J}_i^* \cdot \underline{E} - \underline{J}_c^* \cdot \underline{E} - j\omega \, \underline{B} \cdot \underline{H}^* + j\omega \, \underline{E} \cdot \underline{D}^*.$$

---

[5] Note that ∇ is an operator called real, as it consists of derivations with respect to real variables and real unit vectors, and so if it operates on a real-valued function, then the result also is a real function. It is also said that ∇ commutes with the operation of conjugation:

$$\left(\nabla \times \underline{H}\right)^* = \nabla \times \underline{H}^* = \underline{J}_i^* + \underline{J}_c^* - j\omega \, \underline{D}^*.$$

At this point, we proceed as we did for the analogous theorem in the time domain and then divide by a factor of two, obtaining:

$$\frac{1}{2} \oint_S \underline{n} \cdot (\underline{E} \times \underline{H}^*) \, dS + \frac{1}{2} \int_V \underline{J}_c^* \cdot \underline{E} \, dV + \frac{j\omega}{2} \int_V (\underline{B} \cdot \underline{H}^* - \underline{E} \cdot \underline{D}^*) \, dV$$

$$= -\frac{1}{2} \int_V (\underline{J}_i^* \cdot \underline{E} + \underline{J}_{mi} \cdot \underline{H}^*) \, dV.$$

Let's now impose:

$$\underline{P} = \frac{1}{2} \underline{E} \times \underline{H}^* \qquad p_i = -\frac{1}{2} \underline{J}_i^* \cdot \underline{E}$$

$$p_c = \frac{1}{2} \underline{J}_c^* \cdot \underline{E} \qquad p_{mi} = -\frac{1}{2} \underline{J}_{mi} \cdot \underline{H}^*$$

$$p_H = \frac{j\omega}{2} \underline{B} \cdot \underline{H}^*$$

$$p_E = -\frac{j\omega}{2} \underline{E} \cdot \underline{D}^*$$

obtaining the more compact form:

$$\oint_S \underline{n} \cdot \underline{P} \, dS + \int_V p_c \, dV + \int_V (p_H + p_E) \, dV = \int_V (p_i + p_{mi}) \, dV.$$

The above relation, the complex Poynting's Theorem, is valid both in the case of sinusoidal quantities (phasors) and in the Fourier transform domain. Instead, the interpretation of the theorem which follows is valid only in harmonic regime, because it involves phasors formulas.

We concentrate now, for example, on the term relevant to the flux of the Poynting vector in the time domain $\oint_S \underline{n} \cdot \underline{E} \times \underline{H} \, dS$ and we consider its average value on the period $T = \frac{2\pi}{\omega}$, defining it as usual, $\frac{1}{T} \int_0^T f(t) \, dt = \overline{f(t)}^t$ for a generic function of time $f(t)$. Assuming that the volume V and the surface S are at rest, we can reverse the integrals in space and time, and then carry on the mean operation on the product $\underline{E}(\underline{r}, t) \times \underline{H}(\underline{r}, t)$. Exploiting now the phasor relationship we obtain:

$$\underline{E}(\underline{r}, t) \times \underline{H}(\underline{r}, t) = \text{Re}\left[\underline{E}(\underline{r}) \, e^{j\omega t}\right] \times \frac{1}{2} \left[\underline{H}(\underline{r}) \, e^{j\omega t} + \underline{H}^*(\underline{r}) \, e^{-j\omega t}\right]$$

$$= \text{Re}\left[\frac{1}{2} \underline{E}(\underline{r}) \times \underline{H}(\underline{r}) \, e^{2j\omega t}\right] + \text{Re}\left[\frac{1}{2} \underline{E}(\underline{r}) \times \underline{H}^*(\underline{r})\right],$$

where we put a real quantity (half the sum of two complex conjugates) inside the brackets of the real part and we used the linearity property of the "real part" operator.

Calculating now the average it is obtained:

$$\overline{\underline{E}(\underline{r}, t) \times \underline{H}(\underline{r}, t)}^{t} = \text{Re}\left[\frac{1}{2}\underline{E}(\underline{r}) \times \underline{H}^{*}(\underline{r})\right],$$

as the factor $e^{2j\omega t}$ has zero mean. We then found that the real part of the complex Poynting vector is equal to the average over a period of the Poynting vector in the time domain. Note, however, that the complex Poynting vector isn't the phasor of the corresponding vector in the time domain.

In a similar way one can prove the same result for all the other terms with no derivatives:

$$\overline{-\underline{J}_i(\underline{r}, t) \cdot \underline{E}(\underline{r}, t)}^{t} = \text{Re}\left[-\tfrac{1}{2}\underline{J}_i^{*}(\underline{r}) \cdot \underline{E}(\underline{r})\right],$$

$$\overline{-\underline{J}_{mi}(\underline{r}, t) \cdot \underline{H}(\underline{r}, t)}^{t} = \text{Re}\left[-\tfrac{1}{2}\underline{J}_{mi}(\underline{r}) \cdot \underline{H}^{*}(\underline{r})\right],$$

$$\overline{\underline{J}_c(\underline{r}, t) \cdot \underline{E}(\underline{r}, t)}^{t} = \text{Re}\left[\tfrac{1}{2}\underline{J}_c^{*}(\underline{r}) \cdot \underline{E}(\underline{r})\right].$$

Now we consider the terms with derivatives. For example:

$$\underline{E}(\underline{r}, t) \cdot \frac{\partial \underline{D}(\underline{r}, t)}{\partial t} = \text{Re}\left[\underline{E}(\underline{r}) e^{j\omega t}\right] \cdot \frac{1}{2}\frac{\partial}{\partial t}\left[\underline{D}(\underline{r}) e^{j\omega t} + \underline{D}^{*}(\underline{r}) e^{-j\omega t}\right] =$$

$$= \text{Re}\left[\underline{E}(\underline{r}) e^{j\omega t}\right] \cdot \frac{1}{2} j\omega \left[\underline{D}(\underline{r}) e^{j\omega t} - \underline{D}^{*}(\underline{r}) e^{-j\omega t}\right] =$$

$$= \text{Re}\left[\frac{1}{2} j\omega \underline{E}(\underline{r}) \cdot \underline{D}(\underline{r}) e^{2j\omega t}\right] + \text{Re}\left[-\frac{1}{2} j\omega \underline{E}(\underline{r}) \cdot \underline{D}^{*}(\underline{r})\right],$$

and so:

$$\overline{\underline{E}(\underline{r}, t) \cdot \frac{\partial \underline{D}(\underline{r}, t)}{\partial t}}^{t} = \text{Re}\left[-\frac{1}{2} j\omega \underline{E}(\underline{r}) \cdot \underline{D}^{*}(\underline{r})\right].$$

In a similar way, or in a dual way:

$$\overline{\underline{H}(\underline{r}, t) \cdot \frac{\partial \underline{B}(\underline{r}, t)}{\partial t}}^{t} = \text{Re}\left[-\frac{1}{2} j\omega \underline{H}(\underline{r}) \cdot \underline{B}^{*}(\underline{r})\right] = \text{Re}\left[\frac{1}{2} j\omega \underline{B}(\underline{r}) \cdot \underline{H}^{*}(\underline{r})\right],$$

because complex conjugate numbers have same real part.

We can conclude then that the real part of the equality expressing Poynting's theorem in the frequency domain coincides with the average value on the period of the equality expressing the same theorem in the time domain, and therefore it represents the average value on a period of the power balance in the considered region.

Let us now review some types of materials and let us see the particular expressions obtained for p_c in the frequency domain. In the case of non-dispersive and isotropic medium it is $\underline{J_c} = \sigma \underline{E}$ with σ real, then: $p_c = \frac{1}{2}\sigma \underline{E}^* \cdot \underline{E}$ is a non-negative real quantity, according to its physical meaning of power transferred and not exchanged.[6] In the anisotropic case (still non-dispersive) we have:

$$p_c = \frac{1}{2}\,(\underline{\underline{\sigma}} \cdot \underline{E}^*) \cdot \underline{E}.$$

In order to find the conditions under which this quantity is real, let's consider first the more general case of a dispersive anisotropic medium, in which the tensor $\underline{\underline{\sigma}}$ is in general complex. We then have:

$$p_c = \frac{1}{2}\,(\underline{\underline{\sigma}}^* \cdot \underline{E}^*) \cdot \underline{E}.$$

Assuming p_c real implies that the difference between p_c and its conjugate must be zero:

$$\frac{1}{2}\,(\underline{\underline{\sigma}}^* \cdot \underline{E}^*) \cdot \underline{E} - \frac{1}{2}\,(\underline{\underline{\sigma}} \cdot \underline{E}) \cdot \underline{E}^* = 0,$$

i.e., developing the scalar products:

$$\sum_{i=1}^{3} E_i \sum_{j=1}^{3} \sigma_{ij}^* E_j^* - \sum_{j=1}^{3} E_j^* \sum_{i=1}^{3} \sigma_{ji} E_i = \sum_{i=1}^{3}\sum_{j=1}^{3} (\sigma_{ij}^* - \sigma_{ji})\, E_i\, E_j^* = 0.$$

The previous relation must be valid for any electric field propagating in that medium. Therefore, the only possibility is that $\sigma_{ij}^* = \sigma_{ji}$ i.e. $\underline{\underline{\sigma}}$ is a so-called Hermitian dyadic. In particular, $\underline{\underline{\sigma}}$ becomes real in the non-dispersive case and the above condition coincides with that of symmetry, already verified due to the properties of the crystal lattice. If moreover the medium is passive, then, must also be $p_c > 0$: this means that the matrix has positive eigenvalues.

Let us now consider the power density p_E, starting from the case of the non-dispersive medium, in which it was possible to define a function of state in the time

---

[6] Recall that for real vectors the quantity:

$$A^2 = \underline{A} \cdot \underline{A} = A_x^2 + A_y^2 + A_z^2,$$

can be defined as the square of a vector and so it is defined as positive, and its square root $|\underline{A}| = +\sqrt{\underline{A} \cdot \underline{A}}$ can be assumed as the modulus of the vector. In the case of complex vectors, $A^2 = \underline{A} \cdot \underline{A}$ is not even real in general. To obtain a positive real quantity the modulus is defined as:

$$|\underline{A}| = +\sqrt{\underline{A} \cdot \underline{A}^*} = +\sqrt{A_x A_x^* + A_y A_y^* + A_z A_z^*} = +\sqrt{|A_x|^2 + |A_y|^2 + |A_z|^2}.$$

Often in the literature we talk about "amplitude" instead of the "modulus" although, however, the term amplitude may also sometimes denote a multiplicative complex factor.

domain for the density of electric energy:

$$w_E(\underline{r}, t) = \frac{1}{2} \, \underline{E}(\underline{r}, t) \cdot \underline{D}(\underline{r}, t).$$

Let us now put in the transformed domain:

$$w_E(\underline{r}) = \frac{1}{4} \, \underline{E}(\underline{r}) \cdot \underline{D}^*(\underline{r}) \quad \Rightarrow \quad p_E(\underline{r}) = -2j\omega \, w_E(\underline{r}).$$

Exploiting again the phasor relations we obtain:

$$w_E(\underline{r}, t) = \mathrm{Re}\left[\frac{1}{4} \, \underline{E}(\underline{r}) \cdot \underline{D}(\underline{r}) \, e^{2j\omega t}\right] + \mathrm{Re}\left[\frac{1}{4} \, \underline{E}(\underline{r}) \cdot \underline{D}^*(\underline{r})\right].$$

By averaging with respect to time, we have:

$$\overline{w_E(\underline{r}, t)}^t = \mathrm{Re}[w_E(\underline{r})].$$

Let us now consider the expression of $w_E(\underline{r})$. It was, for an isotropic medium: $\underline{D} = \varepsilon \underline{E}$, with ε real (and positive) independent of ω. Therefore

$$w_E(\underline{r}) = \frac{1}{4} \, \varepsilon \, \underline{E} \cdot \underline{E}^* = \frac{1}{4} \, \varepsilon \, |\underline{E}|^2$$

is a real quantity, so we can omit the real part in the previous relation. This is also a positive definite quantity, in agreement with its physical meaning of energy density. It follows that the power density $p_E(\underline{r})$ is purely imaginary, and therefore:

$$\overline{p_E(\underline{r}, t)}^t = \mathrm{Re}\left[p_E(\underline{r})\right] = 0,$$

according to its physical meaning of exchanged power. As seen in circuit-theory courses, when a complex power is purely imaginary then the active power (which is the real part) is zero, and then the power is all reactive, which means it is stored and exchanged,[7] actually neither carried nor dissipated.

If the medium is anisotropic (still non-dispersive) we have $\underline{D} = \underline{\underline{\varepsilon}} \cdot \underline{E}$ with $\underline{\underline{\varepsilon}}$ real, independent of ω and symmetrical. We then have that:

$$w_E(\underline{r}) = \frac{1}{4} \, \underline{E} \cdot (\underline{\underline{\varepsilon}} \cdot \underline{E}^*)$$

is still a real quantity, in a similar way to what we have previously seen for p_c. It follows again that $p_E(\underline{r})$ is purely imaginary. It will follow that, for physical reasons, $w_E(\underline{r}) > 0$ and this implies positive eigenvalues.

---

[7] Power that goes back and forth in capacitors and inductors.

In the case of dispersive medium, instead, a function of state density of electric energy[8] can no longer be defined. In the isotropic case the constitutive relation was:

$$\underline{D} = \varepsilon(\omega)\,\underline{E} = [\varepsilon_R(\omega) + j\varepsilon_J(\omega)]\underline{E}.$$

This implies, for the power density:

$$p_E(\underline{r}) = -\frac{j\omega}{2}\,\underline{E}\cdot(\varepsilon_R - j\varepsilon_J)\,\underline{E}^* = -\frac{j\omega}{2}\,\varepsilon_R\,\underline{E}\cdot\underline{E}^* - \frac{\omega}{2}\,\varepsilon_J\,\underline{E}\cdot\underline{E}^*,$$

and:

$$\overline{p_E(\underline{r},t)}^{\,t} = \mathrm{Re}[p_E(\underline{r})] = -\frac{\omega}{2}\,\varepsilon_J(\omega)\,\underline{E}\cdot\underline{E}^*.$$

So both active power and dielectric losses are null if $\varepsilon_J(\omega) \equiv 0$. On the other hand, if the medium is dissipative (as it is imposed by Kramers-Krönig relations) and passive, active power will have to be positive and therefore $\varepsilon_J(\omega) < 0$.

In the anisotropic case the power density is:

$$p_E = \frac{j\omega}{2}\,\underline{E}\cdot(\underline{\underline{\varepsilon}}^*\cdot\underline{E}^*).$$

The active power is null[9] if the quantity $\underline{E}\cdot(\underline{\underline{\varepsilon}}^*\cdot\underline{E}^*)$ is real, i.e. if the tensor $\underline{\underline{\varepsilon}}$ is Hermitian. This is demonstrated in a way similar to what we have seen for the tensor $\underline{\underline{\sigma}}$. If instead $\underline{\underline{\varepsilon}}$ is not Hermitian and the medium is passive, the active power will have to be positive.

For duality same considerations apply to p_H, and to the corresponding density of magnetic energy:

$$w_H = \frac{1}{4}\,\underline{B}\cdot\underline{H}^*.$$

In particular the following relation applies to the case of a non-dispersive medium:

$$p_H(\underline{r}) = 2j\omega\,w_H(\underline{r}).$$

In the previous discussion we had to deal with quantities of the following type:

$$\underline{E}\cdot(\underline{\underline{\sigma}}^*\cdot\underline{E}^*) \qquad \underline{E}\cdot(\underline{\underline{\varepsilon}}^*\cdot\underline{E}^*) \qquad \underline{H}^*\cdot(\underline{\underline{\mu}}\cdot\underline{H}).$$

They are, in matrix formalism, structures of the type shown below, e.g. considering the conductivity:

$$(\,\underline{E}\,)\begin{pmatrix}\underline{\underline{\sigma}}^*\end{pmatrix}\begin{pmatrix}\underline{E}^*\end{pmatrix},$$

---

[8] This case corresponds as already seen to the possible presence of dielectric losses.

[9] And so there are no dielectric losses.

which is a homogeneous polynomial of second degree in multiple (complex) variables. But this is just one of the ways to describe a quadratic form of the components of $\underline{E}$. In other words, the matrix uniquely represents the quadratic form.

If the matrix is Hermitian, the quadratic form will be also called Hermitian. In this case, the determinant of the matrix itself and all the eigenvalues are real numbers (recall the general property which states that the determinant is the product of the eigenvalues, each taken with its multiplicity). As we have already seen, if the matrix is Hermitian, whatever the values of the variables are (in our case, the components of the fields), the value assumed by the form is always real, and viceversa.

Furthermore, it is shown that a necessary and sufficient condition for the Hermitian form to be defined positive (which is important for us for physical reasons when we consider the dissipated power and the energy density) is that the (real) eigenvalues of the matrix are all positive.

In the case of dispersive media matrices $\underline{\underline{\varepsilon}}$, $\underline{\underline{\mu}}$ are complex. Note that any complex matrix can be written as follows[10]:

$$\underline{\underline{\varepsilon}} = \underline{\underline{\varepsilon}}' - j\,\underline{\underline{\varepsilon}}'' \quad , \text{ where} \qquad \begin{aligned} \underline{\underline{\varepsilon}}' &= \frac{\underline{\underline{\varepsilon}} + \underline{\underline{\varepsilon}}^{T*}}{2}, \\[2mm] \underline{\underline{\varepsilon}}'' &= j\,\frac{\underline{\underline{\varepsilon}} - \underline{\underline{\varepsilon}}^{T*}}{2}. \end{aligned}$$

as can be easily verified. It is noted that $\underline{\underline{\varepsilon}}'$ and $\underline{\underline{\varepsilon}}''$ are in general complex too, i.e. they are not the real and imaginary parts of $\underline{\underline{\varepsilon}}$. However, this decomposition is important because both $\underline{\underline{\varepsilon}}'$ and $\underline{\underline{\varepsilon}}''$ are Hermitian matrices (whereas the real matrices real part and imaginary part are not generally symmetrical and thus are not Hermitian). Indeed saying that a matrix is Hermitian is equivalent to saying that the matrix coincides with its conjugate transpose (like saying that a matrix is symmetric is equivalent to state that the matrix coincides with its transpose). Taking into account that the transpose of a sum is equal to the sum of the transposes (and the same thing is true for conjugates), the above-mentioned property can be immediately verified. As a Hermitian matrix is in some way a generalization of a real scalar, these two matrices are in a sense generalizations of the real part and the imaginary part (changed in sign) of a scalar. In the case of $\underline{\underline{\varepsilon}}$, $\underline{\underline{\mu}}$ the imaginary part is related (as already seen for the scalar case) to losses phenomena, and must be negative, hence the matrices $\underline{\underline{\varepsilon}}''$ and $\underline{\underline{\mu}}''$ must be positive defined, that is, they need to have positive eigenvalues (assuming a passive medium).

---

[10] With $\underline{\underline{\varepsilon}}^{T*}$ we mean the conjugate transpose of $\underline{\underline{\varepsilon}}$.

1.13 Uniqueness Theorem

The Poynting's theorem just seen can be used to prove the uniqueness theorem, which provides a sufficient condition for the electromagnetic field in order to have a unique solution.

Let us take a region in rest with volume V occupied by a medium (linear, stationary) *non-dispersive*, bounded by a closed surface S.

1.13.1 Uniqueness Theorem in the Time Domain

Consider first the problem in the time domain. Under the following conditions the electromagnetic field is unique in any point of V and at all instants $t > t_o$ where t_o is an initial instant:

1. the field is a solution of Maxwell's equations and satisfies the constitutive relations;
2. for $t = t_o$ the electric field and the magnetic field are assigned at each point of V (initial conditions);
3. for any instant $t > t_o$ either the tangential electric field or the tangential magnetic field is assigned at each point of S (boundary conditions).

Uniqueness theorems are usually proved by indirect proof (reductio ad absurdum), assuming that there are two different electromagnetic fields that both satisfy conditions above; it is then shown that they must necessarily be equal. We will denote the two field solutions with a prime and with a double prime respectively. The impressed currents (field sources) must obviously be the same in both cases. It can then be written:

$$\begin{cases} \nabla \times \underline{E}' = -\underline{J}_{mi} - \dfrac{\partial \underline{B}'}{\partial t} \\ \nabla \times \underline{H}' = \underline{J}_i + \underline{J}'_c + \dfrac{\partial \underline{D}'}{\partial t} \end{cases} , \qquad \begin{cases} \nabla \times \underline{E}'' = -\underline{J}_{mi} - \dfrac{\partial \underline{B}''}{\partial t} \\ \nabla \times \underline{H}'' = \underline{J}_i + \underline{J}''_c + \dfrac{\partial \underline{D}''}{\partial t} \end{cases} .$$

Defining now some "difference" fields and currents:

$$\underline{E}_d = \underline{E}' - \underline{E}'', \ \underline{B}_d = \underline{B}' - \underline{B}'',$$

$$\underline{H}_d = \underline{H}' - \underline{H}'', \ \underline{J}_{cd} = \underline{J}'_c - \underline{J}''_c,$$

$$\underline{D}_d = \underline{D}' - \underline{D}''.$$

Subtracting member to member the Maxwell's equations for the two cases, homogeneous Maxwell's equations are obtained for the difference field:

$$\begin{cases} \nabla \times \underline{E}_d = -\frac{\partial \underline{B}_d}{\partial t} \\ \\ \nabla \times \underline{H}_d = \underline{J}_{cd} + \frac{\partial \underline{D}_d}{\partial t} \end{cases}$$

The difference field also meets the constitutive relations, by virtue of the linearity of the medium, as for example:

$$\underline{D}_d = \underline{D}' - \underline{D}'' = \underline{D}(\underline{E}') - \underline{D}(\underline{E}'') = \underline{D}(\underline{E}' - \underline{E}'') = \underline{D}(\underline{E}_d).$$

Since the Poynting's theorem is a consequence of Maxwell's equations, we can apply it to the difference field in the volume V, taking into account the non-dispersive nature of the medium and the fact that no impressed sources are present here:

$$\oint_S \underline{n} \cdot \underline{E}_d \times \underline{H}_d \, dS + \int_V p_{cd} \, dV + \frac{d}{dt} \int_V \left(W_{E_d} + W_{H_d} \right) dV = 0$$

(taking the time derivative out of the integral, being the volume V at rest).

Let us observe now that the flux of the Poynting vector for the difference field is null for our hypotheses, as on the surface S: either the tangential component of the electric field is assigned, hence $\underline{n} \times \underline{E} \Rightarrow \underline{n} \times \underline{E}' = \underline{n} \times \underline{E}'' \Rightarrow \underline{n} \times \underline{E}_d = 0$ and $\underline{n} \cdot \underline{E}_d \times \underline{H}_d = \underline{n} \times \underline{E}_d \cdot \underline{H}_d = 0$ using the properties of the mixed product according to which we can exchange the scalar and the vector product; or the tangential component of the magnetic field is assigned, hence $\underline{n} \times \underline{H} \Rightarrow \underline{n} \times \underline{H}' = \underline{n} \times \underline{H}'' \Rightarrow \underline{n} \times \underline{H}_d = 0$ and $\underline{n} \cdot \underline{E}_d \times \underline{H}_d = -\underline{n} \cdot \underline{H}_d \times \underline{E}_d = -\underline{n} \times \underline{H}_d \cdot \underline{E}_d = 0$. At the end the following relation remains:

$$\frac{d}{dt} \int_V \left(W_{E_d} + W_{H_d} \right) dV = - \int_V p_{cd} \, dV.$$

Now the two members are integrated in time on the interval $[t_o; t]$, and the internal variable is assumed t' to avoid confusion. We remember that the definite integral of a derivative is equal to the function to be differentiated, calculated in its extremes. Hence:

$$\int_{t_o}^{t} \left[\frac{d}{dt'} \int_V \left(W_{E_d} + W_{H_d} \right) dV \right] dt'$$

$$= \left[\int_V \left(W_{E_d} + W_{H_d} \right) dV \right]_t - \left[\int_V \left(W_{E_d} + W_{H_d} \right) dV \right]_{t_o} = - \int_{t_o}^{t} \left(\int_V p_{cd} \, dV \right) dt'.$$

Let us exploit at this point the initial conditions, not yet used, according to which for $t = t_o$ are assigned $\underline{E}$ and $\underline{H}$ in V $\Rightarrow \underline{E}' \equiv \underline{E}'' \Rightarrow \underline{E}_d \equiv 0$ and $\underline{H}' \equiv \underline{H}'' \Rightarrow \underline{H}_d \equiv 0$ in V $\Rightarrow$ at the initial time $W_{E_d} = W_{H_d} \equiv 0$ in V. For this reason it remains:

$$\left[\int_V (W_{E_d} + W_{H_d}) \, dV \right]_t = - \int_{t_o}^{t} \left(\int_V p_{cd} \, dV \right) dt'.$$

We can observe that W_{E_d} and p_{cd} are positive definite quadratic forms of the components of $\underline{E}_d$ while W_{H_d} is a positive definite quadratic form of the components of $\underline{H}_d$. So the equality written just now is possible only if the two integrals are both zero. Considering the first integral, and this being the integral of the sum of two positive quantities, it follows that both quantities W_{E_d} and W_{H_d} vanish, and so it results $\underline{E}_d = \underline{H}_d \equiv 0 \Rightarrow \underline{E}' \equiv \underline{E}''$ and $\underline{H}' \equiv \underline{H}''$ in each point in V and $\forall t > t_o$. So the theorem is demonstrated, even if the medium is not dissipative ($p_{cd} \equiv 0$ being $\sigma = 0$).

1.13.2 Uniqueness Theorem in the Frequency Domain

We are now going to use the complex Poynting's theorem to derive the uniqueness theorem in the frequency domain. In this case, having eliminated the time dependence, there are no initial conditions but only boundary conditions. Referring to the usual volume V enclosed by a closed surface S and with outward normal unit vector $\underline{n}$, it is unique in any point of V the electromagnetic field which is solution of Maxwell's equations and constitutive relations, and whose is assigned in any point of S either the tangential component of the electric field or the tangential component of the magnetic field. We are also going to assume that the medium inside the volume V is non-dispersive.

Also in this case the theorem is demonstrated by indirect proof (reductio ad absurdum) proceeding in a similar manner as before, so by assuming the existence of two solutions that both satisfy the conditions imposed, and then considering a "difference" electromagnetic field that would be the solution of the homogeneous Maxwell's equations in the frequency domain. Finally, the complex Poynting's theorem is applied to the difference field obtaining:

$$\frac{1}{2} \oint_S \underline{n} \cdot \underline{E}_d \times \underline{H}_d^* \, dS + \int_V p_{cd} \, dV + 2j\omega \int_V (w_{Hd} - w_{Ed}) \, dV = 0.$$

The first integral vanishes due to the boundary conditions, in a similar way to what happened in the time domain, as $\underline{n} \times \underline{H}_d = 0$ implies $\underline{n} \times \underline{H}_d^* = 0$.

There are two integrals remaining, the first of which is purely real, the second purely imaginary (assuming of course that ω is real, that is true in the absence of damping phenomena). It is therefore necessary that both are zero. From the relation:

$$\int_V p_{cd} \, dV = 0$$

in the case of dissipative medium (non-zero conductivity) it follows that, since p_{cd} is a positive definite quadratic form of the components of the difference electric field, such field must be null in the volume V. w_{Ed} is also null, so it remains now:

$$2\omega \int_V w_{Hd}\, dV = 0 \quad \Rightarrow \quad \underline{H}_d \equiv 0 \quad \text{in } V,$$

being w_{Hd} a positive definite quadratic form of the components of the magnetic field. It is ultimately in V:

$$\underline{E}' \equiv \underline{E}'' \ , \quad \underline{H}' \equiv \underline{H}'',$$

and the theorem is finally proved.

In the case of non-dissipative medium (null conductivity), instead, the first integral disappears even if no conditions are imposed on the electric field; since in the second integral there is a difference of quantities depending in general on the angular frequency ω, so it represents an equation in ω which, resolved, provides the particular values of frequency at which the uniqueness theorem is not valid. They are called resonant frequencies of the system, which correspond to particular pairs $\underline{E}_d$, $\underline{H}_d$ not identically zero, representing the electromagnetic field of free oscillations, i.e., not identically zero solutions of the system of homogeneous Maxwell's equations.

Let us finally highlight three points. The first is that the theorem is still valid if the following condition, called impedance condition, is valid in place of the boundary conditions for the tangential components of $\underline{E}$ or $\underline{H}$ on S:

$$\underline{E}_\tau = \zeta_S\, \underline{H}_\tau \times \underline{n},$$

where $\underline{E}_\tau$ and $\underline{H}_\tau$ are tangential fields and ζ_S is a complex quantity, called surface impedance, which is zero in the particular case of perfect conductor and whose real part is positive (for passive media). This boundary condition is commonly used to characterize media between the perfect conductor and perfect dielectric, hence the real media.

The second observation is that when ω is a complex quantity, i.e. in the presence of damping, the theorem is not valid for particular values of ω even for dissipative media. Damped free oscillations appear for these values of ω.

The third observation is that the theorem is also valid for null conductivity medium, if in place of the dissipation due to the Joule effect other types of losses come into play, in particular when dielectric and magnetic losses, related to the dielectric constant and the permeability, appear. The medium is dispersive in such cases and either ε or μ are characterized by an imaginary part with negative sign. In the expression of the uniqueness theorem the following term survives:

$$\frac{1}{2}\, j\omega \int_V (\underline{H}_d^* \cdot \underline{B}_d - \underline{E}_d \cdot \underline{D}_d^*)\, dV,$$

in which it is now possible to separate a real term (related to the imaginary part), which allows us to set to zero separately the electric field (or the magnetic field). It can be concluded therefore that the uniqueness is valid whenever there is some loss mechanism, and then in the more realistic scenarios. Remember, finally, that in the time domain, instead, the uniqueness is always verified.

1.14 Wave Equation

The electromagnetic field can be determined solving the two coupled Maxwell's differential equations of the first order (along with the constitutive relations), but it is usually preferred to solve a second-order differential equation, but in only one vector variable, i.e. one of the two vector fields. The equation obtained is the so-called wave equation and can be considered in both time and frequency domains.

The wave equation in time domain and in the absence of sources is usually shown in the basic physics courses. It is obvious that in practice sources are always present, but they are considered outside the region of interest. Starting from the Maxwell's equations in an isotropic and non-dispersive medium:

$$\begin{cases} \nabla \times \underline{E} = -\mu \dfrac{\partial \underline{H}}{\partial t} \\ \nabla \times \underline{H} = \sigma \underline{E} + \varepsilon \dfrac{\partial \underline{E}}{\partial t} \end{cases}.$$

By applying the curl to the first equation we obtain:

$$\nabla \times \nabla \times \underline{E} = \nabla \nabla \cdot \underline{E} - \nabla^2 \underline{E} = -\nabla \times \left(\mu \dfrac{\partial \underline{H}}{\partial t} \right).$$

Let us suppose now that the medium be homogeneous (this is a hypothesis always necessary in order to derive the wave equation), and let us apply the Schwarz theorem, obtaining:

$$\nabla \nabla \cdot \underline{E} - \nabla^2 \underline{E} = -\mu \dfrac{\partial}{\partial t} (\nabla \times \underline{H}).$$

Now, by inserting the second Maxwell's equation, we get:

$$\nabla \nabla \cdot \underline{E} - \nabla^2 \underline{E} = -\mu \dfrac{\partial}{\partial t} \left(\sigma \underline{E} + \varepsilon \dfrac{\partial \underline{E}}{\partial t} \right).$$

On the other hand (again in the assumption of homogeneity of the medium):

$$\nabla \cdot \underline{E} = \dfrac{\rho}{\varepsilon}.$$

Then:

$$\nabla^2 \underline{E} - \mu \sigma \dfrac{\partial \underline{E}}{\partial t} - \mu \varepsilon \dfrac{\partial^2 \underline{E}}{\partial t^2} = \dfrac{\nabla \rho}{\varepsilon}.$$

Assuming the absence of free charges the following homogeneous equation is obtained:

$$\nabla^2 \underline{E} - \mu\sigma \frac{\partial \underline{E}}{\partial t} - \mu\varepsilon \frac{\partial^2 \underline{E}}{\partial t^2} = 0.$$

The well-known d'Alembert equation follows in the particular case of non-dissipative medium ($\sigma = 0$):

$$\nabla^2 \underline{E} - \mu\varepsilon \frac{\partial^2 \underline{E}}{\partial t^2} = \nabla^2 \underline{E} - \frac{1}{v^2} \frac{\partial^2 \underline{E}}{\partial t^2} = 0,$$

having introduced the light speed in the medium:

$$v = \frac{1}{\sqrt{\mu\varepsilon}}.$$

Sometimes is also used the so-called d'Alembert operator:

$$\square = \nabla^2 - \frac{1}{v^2} \frac{\partial^2}{\partial t^2},$$

so that the equation becomes simply:

$$\square \underline{E} = 0.$$

Operating the Fourier transform (with respect to time) of the d'Alembert equation, the following is obtained:

$$\nabla^2 \underline{E} + \omega^2 \mu\varepsilon \underline{E} = \nabla^2 \underline{E} + k^2 \underline{E} = 0,$$

with $k^2 = \omega^2 \mu\varepsilon$, being $k = \omega\sqrt{\mu\varepsilon} = \frac{\omega}{v}$ the so-called wave number (in the medium) or the propagation constant, or the propagation wave number in the medium. This equation is the well-known Helmholtz equation and we will see shortly how its validity is quite general, as it is valid even for dispersive (in time) media.

The wave number is related to the wavelength λ of the medium by the relation:

$$k = \frac{2\pi}{\lambda}.$$

There is also another fundamental relation:

$$v = \lambda f.$$

Note that λ plays the role of the spatial period of the field, while k is a kind of space angular frequency.

In a vacuum, in particular:

$$k_o = \frac{\omega}{c} = \frac{2\pi}{\lambda_o},$$

being[11]:

$$c = \frac{1}{\sqrt{\mu_o \varepsilon_o}} \quad , \quad v = \frac{c}{\sqrt{\mu_r \varepsilon_r}} \quad , \quad \lambda = \frac{\lambda_o}{\sqrt{\mu_r \varepsilon_r}}.$$

This means that in a material medium it is like if the field is "compressed" with respect to a vacuum, since the wavelength decreases.[12] Very often we can assume $\mu_r = 1$ in the media that we consider in practice. The frequency instead does not depend on the medium.

1.14.1 Helmholtz Equation

We derive now the non-homogeneous Helmholtz equation starting from the Maxwell's equations in the transformed domain. Let us consider again a homogeneous, isotropic, but also generally dispersive medium

$$\begin{cases} \nabla \times \underline{E} = -\underline{J}_{mi} - j\omega\mu\,\underline{H} \\ \nabla \times \underline{H} = \underline{J}_i + \sigma\,\underline{E} + j\omega\varepsilon\,\underline{E} = \underline{J}_i + (\sigma + j\omega\varepsilon)\underline{E} \end{cases}.$$

For compactness of notation it is usually put:

$$j\omega\varepsilon_c = \sigma + j\omega\varepsilon \quad \Rightarrow \quad \nabla \times \underline{H} = \underline{J}_i + j\omega\varepsilon_c\,\underline{E},$$

with:

$$\varepsilon_c = \varepsilon + \frac{\sigma}{j\omega} = \varepsilon - j\frac{\sigma}{\omega}$$

equivalent dielectric constant (complex in general). Note that the duality principle is still valid for the new pair of equations using the transformations:

$$\varepsilon_c \to \mu \qquad \mu \to \varepsilon_c.$$

Taking the divergences we have, in the usual assumption of homogeneous medium:

$$0 = -\nabla \cdot \underline{J}_{mi} - j\omega\mu\,\nabla \cdot \underline{H} \quad \Rightarrow \quad \nabla \cdot \underline{H} = -\frac{\nabla \cdot \underline{J}_{mi}}{j\omega\mu},$$

$$0 = \nabla \cdot \underline{J}_i + j\omega\varepsilon_c\,\nabla \cdot \underline{E} \quad \Rightarrow \quad \nabla \cdot \underline{E} = -\frac{\nabla \cdot \underline{J}_i}{j\omega\varepsilon_c}.$$

---

[11] Then the light in material media is slower than in a vacuum.

[12] So, for example, using dielectrics with high ε_r we can miniaturize components, such as for example resonators.

At this point we apply the curl to the first equation and then we substitute the second:

$$\nabla \times \nabla \times \underline{E} = \nabla \nabla \cdot \underline{E} - \nabla^2 \underline{E} = -\nabla \times \underline{J}_{mi} - j\omega\mu \, (\underline{J}_i + j\omega\varepsilon_c \, \underline{E}).$$

By substituting the obtained expression for the divergence of $\underline{E}$:

$$\nabla^2 \underline{E} + k^2 \underline{E} = \nabla \times \underline{J}_{mi} + j\omega\mu \, \underline{J}_i - \frac{\nabla \nabla \cdot \underline{J}_i}{j\omega\varepsilon_c},$$

having placed $k^2 = \omega^2 \mu \varepsilon_c$.

The above equation has a source term which is rather complicated (in particular, it also requires the derivability of impressed sources). Applying again the duality principle, taking the curl of the second Maxwell's equation and substituting the first one, we obtain the equation for the magnetic field:

$$\nabla^2 \underline{H} + k^2 \underline{H} = -\nabla \times \underline{J}_i + j\omega\varepsilon_c \, \underline{J}_{mi} - \frac{\nabla \nabla \cdot \underline{J}_{mi}}{j\omega\mu}.$$

Note that a solution of the Helmholtz equation is not necessarily a solution of Maxwell's equations, as the condition on the divergence also must be imposed. Once one of the two vector fields is found, the other one can be obtained from the curl Maxwell's equation of the known vector field.

Finally, note that Fourier transforming the equation:

$$\nabla^2 E - \mu\sigma \, \frac{\partial \underline{E}}{\partial t} - \mu\varepsilon \, \frac{\partial^2 \underline{E}}{\partial t^2} = 0$$

it is obtained:

$$\nabla^2 \underline{E} - j\omega\mu\sigma \, \underline{E} + \omega^2 \mu\varepsilon \, \underline{E} = 0 \quad \Rightarrow \quad \nabla^2 \underline{E} + \omega^2 \mu \left(\varepsilon - j\frac{\sigma}{\omega} \right) \underline{E} = 0,$$

from which, replacing ε_c:

$$\nabla^2 \underline{E} + k^2 \, \underline{E} = 0.$$

1.15 Electromagnetic Potentials

It is sometimes useful to solve the electromagnetic problem, not directly in terms of electromagnetic fields, but making use of auxiliary scalar and vector functions (in the same way as already seen in electrostatics and magnetostatics), representing the electromagnetic field through such functions which take the name of electrodynamic or electromagnetic potentials. There are different kinds of potentials that can be considered. We will present the most commonly used: in order to do so, we firstly need, assuming the validity of linearity, to apply the principle of superposition of effects to the non-homogeneous Maxwell's equations. Then we suppose firstly the absence of the impressed magnetic currents, then the absence of impressed electrical currents, and finally we sum the two results in order to deal with the general case.

Let's start from the frequency domain. By imposing $\underline{J}_{mi} \equiv 0$ we get:

$$\begin{cases} \nabla \times \underline{E} = -j\omega\mu\,\underline{H} \\ \nabla \times \underline{H} = \underline{J}_i + j\omega\varepsilon_c\,\underline{E} \end{cases}$$

Assuming a homogeneous medium, and applying the divergence to the first equation, it is obtained:

$$\nabla \cdot \underline{H} = 0,$$

in the assumption that the domain is simple surface connected, it can be imposed:

$$\underline{H} = \nabla \times \underline{A},$$

where $\underline{A}$ is the vector potential, defined apart from the gradient of a scalar function Φ, i.e. assuming:

$$\underline{A}' = \underline{A} + \nabla\Phi \Rightarrow \underline{H} = \nabla \times \underline{A}'.$$

The non-uniqueness of the vector field $\underline{A}$ derives from the fact that its curl is assigned while its divergence is not. There is in fact a theorem of vector analysis, the Helmholtz theorem, which states that a vector field is uniquely determined when both curl and divergence are assigned.

Substituting $\underline{H} = \nabla \times \underline{A}$ in the first Maxwell's equation it is obtained:

$$\nabla \times \underline{E} = -j\omega\mu\,\nabla \times \underline{A} \quad \Rightarrow \quad \nabla \times (\underline{E} + j\omega\mu\,\underline{A}) = 0,$$

from which, if the domain is simple linear connected, it follows:

$$\underline{E} + j\omega\mu\,\underline{A} = -\nabla V \quad \Rightarrow \quad \underline{E} = -j\omega\mu\,\underline{A} - \nabla V.$$

The quantity V is said scalar potential. We therefore expressed the electromagnetic field in terms of a pair of potentials. This is not the only possible choice: for example, the so-called Hertz potentials can also be used.

Note now that if one changes the vector potential according to the transformation $\underline{A}' = \underline{A} + \nabla\Phi$ (which leaves unchanged the magnetic field, as it should be) then one has to change the scalar potential in order to maintain the electric field unchanged too. In fact it follows:

$$\underline{E} = -j\omega\mu\,(\underline{A}' - \nabla\Phi) - \nabla V = -j\omega\mu\,\underline{A}' + j\omega\mu\,\nabla\Phi - \nabla V =$$

$$= -j\omega\mu\,\underline{A}' - \nabla(V - j\omega\mu\,\Phi) = -j\omega\mu\,\underline{A}' - \nabla V'$$

having put $V' = V - j\omega\mu\,\Phi$. The transformation:

$$\begin{cases} \underline{A}' = \underline{A} + \nabla\Phi \\ V' = V - j\omega\mu\,\Phi \end{cases}$$

is called *gauge* transformation, where Φ is an arbitrary scalar function.

At this point, the problem moves to the evaluation of the pair $\underline{A}$, V. In order to reach our objective, we insert $\underline{H} = \nabla\times\underline{A}$ in the second Maxwell's equation, and so we have:

$$\nabla\times\nabla\times\underline{A} = \underline{J_i} + j\omega\varepsilon_c\,\underline{E},$$

from which:

$$\nabla\nabla\cdot\underline{A} - \nabla^2\underline{A} = \underline{J_i} + j\omega\varepsilon_c\,(-j\omega\mu\,\underline{A} - \nabla V) = \underline{J_i} + \omega^2\mu\varepsilon_c\,\underline{A} - j\omega\varepsilon_c\,\nabla V.$$

The above equation, in which both potentials are present, can be much simplified if $\underline{A}$ and V satisfy the Lorenz condition (or gauge):

$$\nabla\cdot\underline{A} = -j\omega\varepsilon_c\,V,$$

(note that it is not the only choice: for example, also the so-called Coulomb condition could be applied) for which we obtain the non-homogeneous Helmholtz equation in one unknown variable $\underline{A}$:

$$\nabla^2\underline{A} + k^2\underline{A} = -\underline{J_i}.$$

As shown, the source term of this equation is much simpler than the one for the corresponding equation for $\underline{E}$ (or for $\underline{H}$), and in particular it does not require differentiability of the impressed sources. This equation is solved by means of the so-called Green's function.

Let us suppose now that we have a pair $\underline{A}_o$, V_o of potentials that do not meet the Lorenz condition. It is then possible, using the gauge transformations, to get a new pair $\underline{A}$, V that satisfies it, through an appropriate choice of the function Φ. On the other hand this means, with reference again to the Helmholtz theorem, that is assigned the divergence of $\underline{A}$, whose curl was already assigned. Starting from:

$$\begin{aligned} \underline{A} &= \underline{A}_o + \nabla\Phi, \\ V &= V_o - j\omega\mu\,\Phi, \end{aligned}$$

it will now be:

$$\nabla\cdot(\underline{A}_o + \nabla\Phi) = -j\omega\varepsilon_c\,(V_o - j\omega\mu\,\Phi)$$

$$\Rightarrow \quad \nabla\cdot\underline{A}_o + \nabla\cdot\nabla\Phi = -j\omega\varepsilon_c\,V_o - \omega^2\mu\varepsilon_c\,\Phi$$

$$\Rightarrow \quad \nabla^2\Phi + k^2\,\Phi = -\nabla\cdot\underline{A}_o - j\omega\varepsilon_c\,V_o,$$

which is still a non-homogeneous Helmholtz equation in the variable Φ, since the known term (source) is different from zero by assumption.

Once the Lorenz condition is met, the scalar potential can be expressed as a function of the vector potential:

$$V = \frac{\nabla \cdot \underline{A}}{-j\omega\varepsilon_c},$$

and we can then express the electric field in terms of the only variable $\underline{A}$:

$$\underline{E} = -j\omega\mu\,\underline{A} + \frac{\nabla\nabla \cdot \underline{A}}{j\omega\varepsilon_c}.$$

Let us consider finally the dual case in which $\underline{J_j} \equiv 0$ and $\underline{J_{mi}} \neq 0$. Then, applying the divergence to the second Maxwell's equation it follows, again assuming the homogeneity of the medium:

$$\nabla \cdot \underline{E} = 0 \quad \Rightarrow \quad \underline{E} = -\nabla \times \underline{F}$$

where the minus sign has been introduced because $\underline{F}$ results the dual quantity of $\underline{A}$, having:

$$\underline{A} \to \underline{F} \quad , \quad \underline{F} \to -\underline{A}.$$

The equation for $\underline{H}$ is obtained again by using the duality principle:

$$\underline{H} = -j\omega\varepsilon_c\,\underline{F} - \nabla U = -j\omega\varepsilon_c\,\underline{F} + \frac{\nabla\nabla \cdot \underline{F}}{j\omega\mu},$$

being the scalar potential U the dual of V. We have:

$$V \to U \quad , \quad U \to -V.$$

The equation to be solved in this case is:

$$\nabla^2 \underline{F} + k^2 \underline{F} = -\underline{J_{mi}}.$$

Let us mention now the corresponding procedure in the time domain. Let us suppose, for simplicity, that the medium is isotropic non-dispersive and non-dissipative. In the case $\underline{J_{mi}} = 0$, it follows:

$$\begin{cases} \nabla \times \underline{E} = -\mu \dfrac{\partial H}{\partial t} \\[2mm] \nabla \times \underline{H} = \underline{J_j} + \varepsilon \dfrac{\partial E}{\partial t} \end{cases}.$$

Applying again the divergence to the first equation for homogeneous media and then using the Schwarz theorem, we obtain:

$$\frac{\partial}{\partial t}(\nabla \cdot \underline{H}) = 0$$

$$\Rightarrow \quad \nabla \cdot \underline{H} = constant \quad \Rightarrow \quad \nabla \cdot \underline{H} = 0 \quad \Rightarrow \quad \underline{H} = \nabla \times \underline{A}$$

$$\Rightarrow \quad \nabla \times \left(\underline{E} + \mu \frac{\partial \underline{A}}{\partial t} \right) = 0 \quad \Rightarrow \quad \underline{E} = -\mu \frac{\partial \underline{A}}{\partial t} - \nabla V.$$

Then, from the second equation:

$$\nabla \times \nabla \times \underline{A} = \nabla \nabla \cdot \underline{A} - \nabla^2 \underline{A} = \underline{J}_j + \varepsilon \frac{\partial}{\partial t} \left(-\mu \frac{\partial \underline{A}}{\partial t} - \nabla V \right)$$

$$= \underline{J}_j - \mu \varepsilon \frac{\partial^2 \underline{A}}{\partial t^2} - \varepsilon \nabla \frac{\partial V}{\partial t}.$$

By imposing the Lorenz condition:

$$\nabla \cdot \underline{A} = -\varepsilon \frac{\partial V}{\partial t}$$

follows the non-homogeneous d'Alembert equation for the vector potential:

$$\nabla^2 \underline{A} - \mu \varepsilon \frac{\partial^2 \underline{A}}{\partial t^2} = -\underline{J}_j.$$

Once we have solved this equation, we obtain $\underline{A}$, then $\underline{H}$ and finally $\underline{E}$ by the Lorenz[13] condition, which yields:

$$V = -\frac{1}{\varepsilon} \int_{t_o}^{t} \nabla \cdot \underline{A} \, dt'$$

$$\Rightarrow \quad \underline{E} = -\mu \frac{\partial \underline{A}}{\partial t} + \frac{1}{\varepsilon} \int_{t_o}^{t} \nabla \nabla \cdot \underline{A} \, dt'.$$

The other case follows from duality.

---

[13] Note that in frequency domain the corresponding formula shows a division by $j\omega$ instead of an integration with respect to t.

Chapter 2
Properties of Plane Electromagnetic Waves

Abstract After introducing some general features of wave functions (equiphase and equi-amplitude surfaces, phase vector, phase velocity), the treatment is particularized to the fundamental (for theory and applications) case of plane waves: the general properties and the various wave types are reviewed. The important concept of plane-wave spectrum is investigated. Non-monochromatic fields and the concept of group velocity are considered. Moreover, the fundamental reflection and transmission properties at plane interfaces are examined in detail, for normal and oblique incidence, horizontal and vertical polarization. Fresnel coefficients are derived, total reflection and total transmission are explained, good conductors are included.

2.1 Wave Functions

Let us consider now the homogeneous Helmholtz equation for a generic vector wave function $\underline{A}$ (not necessarily an electromagnetic field):

$$\nabla^2 \underline{A} + k^2 \underline{A} = 0 .$$

We can project this equation on the three Cartesian axes x, y, z recalling that, in these coordinate system, the components of the vector Laplacian are the scalar Laplacian of the components, and so we obtain, for example, in the x direction:

$$\nabla^2 A_x + k^2 A_x = 0 .$$

Therefore we can refer to scalar wave functions $A(x, y, z)$ satisfying the $\nabla^2 A + k^2 A = 0$.

The generic scalar complex wave function can always be written as follows:

$$A(x, y, z) = M(x, y, z) \, e^{-j \, \Phi(x,y,z)} ,$$

having highlighted magnitude and phase. We are usually interested in phase changes rather than in the phase value: in other words, the function Φ is typically defined apart from an additive constant.

© Springer International Publishing Switzerland 2015
F. Frezza, *A Primer on Electromagnetic Fields*,
DOI 10.1007/978-3-319-16574-5_2

In the particular case in which $\Phi(x, y, z) = constant$ in a certain region of space, the wave is said to be stationary in that region. When this is not the case, the wave is said to be progressive and the previous relation, for different values of the constant, defines a family of surfaces which are called equiphase surfaces. The particular shape of these surfaces (e.g. planes, cylinders, spheres) is used to name the wave (plane, cylindrical, spherical).

The *phase vector* is defined as $\beta = \nabla\Phi$. It can be shown that, as for any gradient, its direction is the direction in which the phase variation is maximum; moreover $\underline{\beta}$ is orthogonal to equiphase surfaces; finally, obviously, $\underline{\beta} \equiv 0$ for a standing wave.

As it should be known, the superposition of two progressive waves having same module and of opposite phases generates a standing wave. To show this, let us write:

$$A_1(x, y, z) = M(x, y, z)\,e^{-j\,\Phi(x,y,z)}\,,$$
$$A_2(x, y, z) = M(x, y, z)\,e^{j\,\Phi(x,y,z)}\,,$$

it follows:

$$A_1(x, y, z) + A_2(x, y, z) = M(x, y, z)\left[e^{j\,\Phi(x,y,z)} + e^{-j\,\Phi(x,y,z)}\right]$$

$$= 2\,M(x, y, z)\,\cos\left[\Phi(x, y, z)\right],$$

which is a real function. Therefore, its phase is zero (or π) in the whole space and so the wave is stationary.

The equation $M(x, y, z) = constant$ defines, if not identically verified, a family of surfaces which are called equi-amplitude surfaces. A wave is said to be uniform (more often referred in the literature as homogeneous) if the amplitude is constant on the equiphase surfaces. This occurs, either when the previous relation is identically verified, or when the equi-amplitude surfaces coincide with equiphase surfaces.

Returning now to the time domain, in monochromatic regime we have:

$$A(x, y, z, t) = \mathrm{Re}\left[A(x, y, z)\,e^{j\omega t}\right] = M(x, y, z)\,\cos\left[\omega t - \Phi(x, y, z)\right].$$

The function $\Psi(x, y, z, t) = \omega t - \Phi(x, y, z)$ represents the phase variation in space and time. The phase velocity v_r in the direction of the unit vector $\underline{r}_o$ is defined as the speed of a hypothetical observer which moves in that direction in such a way that he does not observe phase variations. For such observer therefore it is:

$$d\Psi = 0 \quad \Rightarrow \quad \omega\,dt - d\Phi = 0$$
$$\Rightarrow \omega\,dt - \frac{\partial\Phi}{\partial r}dr = \omega\,dt - \nabla\Phi\cdot\underline{r}_o dr =$$
$$= \omega\,dt - \underline{\beta}\cdot\underline{r}_o dr = \omega\,dt - \beta_r dr = 0$$
$$\Rightarrow v_r = \frac{dr}{dt} = \frac{\omega}{\beta_r}\,.$$

Note that a phase-velocity vector $\underline{v}$ is not defined: the phase velocity is essentially a scalar quantity, i.e. v_r can not be the component of a hypothetical vector $\underline{v}$ along the direction $\underline{r}_0$. In fact, the existence of such a vector should imply:

$$\underline{v} = \underline{x}_o \frac{\omega}{\beta_x} + \underline{y}_o \frac{\omega}{\beta_y} + \underline{z}_o \frac{\omega}{\beta_z},$$

with $\underline{v} \cdot \underline{r}_o \neq \dfrac{\omega}{\underline{\beta} \cdot \underline{r}_o} = v_r$.

2.2 Plane Waves

We will study now a particular solution of the homogeneous Helmholtz equation in free space (i.e., no discontinuity surfaces) in a homogeneous, isotropic, possibly dispersive medium. The equation for the electric field is the following:

$$\nabla^2 \underline{E} + k^2 \underline{E} = 0.$$

We proceed now according to the technique of the so-called separation of variables considering solutions of the form (in Cartesian coordinates):

$$\underline{E}(x, y, z) = \underline{E}_o X(x) Y(y) Z(z),$$

being $\underline{E}_o$ a constant vector (in general complex) and X, Y, Z three scalar functions (in general complex). Let us introduce this form of solution in the equation above, obtaining[1]:

$$\underline{E}_o \left(\frac{d^2 X}{dx^2} YZ + X \frac{d^2 Y}{dy^2} Z + XY \frac{d^2 Z}{dz^2} \right) + k^2 \underline{E}_o XYZ = 0.$$

Collecting $\underline{E}_o$ as a common factor (being this certainly different from zero, otherwise only the trivial solution identically zero, obviously always present in homogeneous equations, would be obtained) and applying the zero-product property, it follows:

$$\frac{d^2 X}{dx^2} YZ + X \frac{d^2 Y}{dy^2} Z + XY \frac{d^2 Z}{dz^2} + k^2 XYZ = 0.$$

Assuming $XYZ \neq 0$ and dividing everything by this quantity:

$$\frac{1}{X} \frac{d^2 X}{dx^2} + \frac{1}{Y} \frac{d^2 Y}{dy^2} + \frac{1}{Z} \frac{d^2 Z}{dz^2} + k^2 = 0,$$

---

[1] Note that $\nabla^2 \underline{E} = \underline{E}_o \nabla^2 (XYZ)$.

where the first term depends only on x, the second only on y, the third only on z, while the fourth is constant.

Differentiating with respect to x, then we obtain:

$$\frac{d}{dx}\left[\frac{1}{X}\frac{d^2X}{dx^2}\right] = 0 \quad \Rightarrow \quad \frac{1}{X}\frac{d^2X}{dx^2} = constant = -k_x^2 ,$$

k_x being a generic complex constant. Similarly, differentiating with respect to y and with respect to z, we obtain:

$$\frac{1}{Y}\frac{d^2Y}{dy^2} = constant = -k_y^2 , \quad \frac{1}{Z}\frac{d^2Z}{dz^2} = constant = -k_z^2 .$$

i.e. three equations of harmonic motions. The three constants in the above equations are not independent, since it must be:

$$-k_x^2 - k_y^2 - k_z^2 + k^2 = 0 \quad \Rightarrow \quad k^2 = k_x^2 + k_y^2 + k_z^2 .$$

The previous relation, called separability condition, should be imposed to ensure that the three equations of harmonic motions are equivalent to the initial equation. In particular, the three constants k_x, k_y and k_z can not be simultaneously zero, being $k^2 = \omega^2 \mu \varepsilon_c \neq 0$.

Let us now examine the first equation:

$$\frac{d^2X}{dx^2} + k_x^2 X = 0 .$$

The general solution is obtained considering separately the two cases $k_x \neq 0$ and $k_x = 0$. We have in the first case:

$$X(x) = X_o^+ e^{-jk_x x} + X_o^- e^{jk_x x} ,$$

with X_o^+ and X_o^- arbitrary complex constants. We could alternatively write the solution in terms of sinusoidal functions, but the actual choice is more convenient to describe progressive waves in free space, in the absence of obstacles that can generate reflections and standing waves. In particular, the first term represents a wave that propagates in the direction of the positive x, while the second one is the wave that propagates in the direction of negative x, as highlighted by using the apexes "+" and "−". Assuming $k_x = 0$, we have instead:

$$X(x) = X_{01} x + X_{02} ,$$

where X_{01} and X_{02} are generic complex constants. Analogous expressions are obtained for $Y(y)$ and $Z(z)$.

Now we are going to specialise further our particular solution by assuming $X_o^- = 0$ when $k_x \neq 0$ (i.e. we are going to consider only terms which propagate in the direction of the positive x) and, analogously, $Y_o^- = Z_o^- = 0$. On the other hand if $k_x = 0$ we assume $X_{01} = 0$, so that $X(x) = constant$, and this constant can be included in the other equation, described by the exponential term, allowing it to cover the case $k_x = 0$. In the same way we put $Y_{01} = Z_{01} = 0$. Finally, joining all parts together and embedding all the remaining X_o^+, Y_o^+ and Z_o^+ constants in the constant vector $\underline{E}_o$, we have:

$$\underline{E}(x, y, z) = \underline{E}_o \, e^{-j(k_x x + k_y y + k_z z)} .$$

There are two parts in the above expression. The vector factor (constant) determines the characteristics of polarization, while the exponential scalar factor (dependent on the coordinates) determines the propagation characteristics.

Defining the complex propagation vector:

$$\underline{k} = k_x \, \underline{x}_o + k_y \, \underline{y}_o + k_z \, \underline{z}_o ,$$

and recalling that:

$$\underline{r} = x \, \underline{x}_o + y \, \underline{y}_o + z \, \underline{z}_o ,$$

we have:

$$k_x \, x + k_y \, y + k_z \, z = \underline{k} \cdot \underline{r} .$$

So the electric field ultimately assumes the following expression:

$$\underline{E}(x, y, z) = \underline{E}(\underline{r}) = \underline{E}_o \, e^{-j\underline{k} \cdot \underline{r}} .$$

There are two conditions that need to be added to this solution in order to be valid: the condition of separability[2] since we are describing a wave, i.e. a solution of the Helmholtz equation; and the condition $\nabla \cdot \underline{E} = 0$, which was used to derive the Helmholtz equation from Maxwell's equations.

It is important to point out that for a functional form of the type $\underline{E} = \underline{E}_o \, e^{-j\underline{k} \cdot \underline{r}}$ the operator ∇ coincides with the vector $-j\underline{k}$: this is due to the form assumed in this case by the spatial derivatives. In fact, we have:

$$\nabla = \underline{x}_o(-jk_x) + \underline{y}_o(-jk_y) + \underline{z}_o(-jk_z) = -j\underline{k} ,$$

from which

$$\nabla(e^{-j\underline{k} \cdot \underline{r}}) = -j\underline{k} \, e^{-j\underline{k} \cdot \underline{r}} ,$$

$$\nabla \cdot \left(\underline{E}_o \, e^{-j\underline{k} \cdot \underline{r}} \right) = -j\underline{k} \cdot \underline{E}_o \, e^{-j\underline{k} \cdot \underline{r}} ,$$

---

[2] $\quad \underline{k} \cdot \underline{k} = k_x^2 + k_y^2 + k_z^2 = \omega^2 \mu \varepsilon_c .$

$$\nabla \times \left(\underline{E}_o \, e^{-j\underline{k}\cdot\underline{r}} \right) = -j\underline{k} \, \times \, \underline{E}_o \, e^{-j\underline{k}\cdot\underline{r}},$$

and moreover:

$$\nabla^2 \, e^{-j\underline{k}\cdot\underline{r}} = \nabla\cdot\nabla\left(e^{-j\underline{k}\cdot\underline{r}}\right) = -\underline{k}\cdot\underline{k} \, e^{-j\underline{k}\cdot\underline{r}}$$

$$\Rightarrow (\nabla^2 + k^2) \, e^{-j\underline{k}\cdot\underline{r}} = 0 \quad \text{if} \quad \underline{k}\cdot\underline{k} = k^2,$$

as it should be. And finally:

$$\nabla^2 \, (\underline{E}_o \, e^{-j\underline{k}\cdot\underline{r}}) = \nabla\cdot\nabla(\underline{E}_o \, e^{-j\underline{k}\cdot\underline{r}}) =$$

$$= -\underline{k}\cdot(\underline{k} \, \underline{E}_o) \, e^{-j\underline{k}\cdot\underline{r}} = -(\underline{k}\cdot\underline{k}) \, \underline{E}_o \, e^{-j\underline{k}\cdot\underline{r}}$$

$$\Rightarrow (\nabla^2 + k^2)(\underline{E}_o \, e^{-j\underline{k}\cdot\underline{r}}) = 0 \quad \text{if} \quad \underline{k}\cdot\underline{k} = k^2 \, .$$

as it should be.

We can now come back to the condition $\nabla\cdot\underline{E} = 0$, which, from what we have seen, assumes the following form:

$$-j\underline{k}\cdot\underline{E}_o \, e^{-j\underline{k}\cdot\underline{r}} = 0 \, .$$

The value of the exponential term is always different from 0, even in the complex field, so the zero-product property implies:

$$\underline{k}\cdot\underline{E}_o = 0 \, .$$

This is the condition for which our plane wave is also solution of Maxwell's equations.

Separating the real and the imaginary parts of the $\underline{k}$ components, we have:

$$k_x = \beta_x - j\alpha_x \, ,$$
$$k_y = \beta_y - j\alpha_y \, ,$$
$$k_z = \beta_z - j\alpha_z \, ,$$

and by defining the two real vectors:

$$\underline{\beta} = \underline{x}_o \, \beta_x + \underline{y}_o \, \beta_y + \underline{z}_o \, \beta_z \, ,$$
$$\underline{\alpha} = \underline{x}_o \, \alpha_x + \underline{y}_o \, \alpha_y + \underline{z}_o \, \alpha_z \, ,$$

which are named respectively *phase vector* and *attenuation vector*, we have:

$$\underline{E}(x, y, z) = \underline{E}(\underline{r}) = \underline{E}_o \, e^{-j\underline{\beta}\cdot\underline{r}} \, e^{-\underline{\alpha}\cdot\underline{r}} \, .$$

Let us now consider the case of a general vector wave function to acquire a better understanding on the meaning of the vectors $\underline{\beta}$ and $\underline{\alpha}$. If the three components have the same phase (apart from a constant), we can generalize directly and unequivocally the concepts of equiphase surface, phase vector and phase velocity (in sinusoidal regime). In addition, if the three components present a common amplitude factor (containing the dependence on coordinates), it is possible to extend the concept of equi-amplitude surface and uniform wave. Well, this is exactly the case of our plane waves in free space. There is a common phase factor $e^{-j\underline{\beta}\cdot\underline{r}}$, so:

$$\Phi(x, y, z) = \underline{\beta}\cdot\underline{r} = \beta_x\, x + \beta_y\, y + \beta_z\, z\,.$$

Therefore the phase vector for a generic wave

$$\nabla\Phi = \underline{x}_o\frac{\partial\Phi}{\partial x} + \underline{y}_o\frac{\partial\Phi}{\partial y} + \underline{z}_o\frac{\partial\Phi}{\partial z} = \beta_x\,\underline{x}_o + \beta_y\,\underline{y}_o + \beta_z\,\underline{z}_o$$

coincides with the $\underline{\beta}$. Moreover, there is a common amplitude factor $e^{-\underline{\alpha}\cdot\underline{r}}$.

In order to determine the equiphase surfaces, let us observe that if the two points P and P', identified by the vectors $\underline{r}$ and $\underline{r}'$, belong to an equiphase surface, then it must be $\Phi(\underline{r}) = \Phi(\underline{r}')$, and therefore:

$$\underline{\beta}\cdot\underline{r} = \underline{\beta}\cdot\underline{r}' \qquad \Rightarrow \qquad \underline{\beta}\cdot(\underline{r} - \underline{r}') = 0\,.$$

It follows that the vector $\underline{r} - \underline{r}' = \overrightarrow{P'P}$ must be orthogonal to $\underline{\beta}$, i.c. it must lie on a plane orthogonal to $\underline{\beta}$. We conclude that the equiphase surfaces are planes normal to $\underline{\beta}$ and defined by the equation $\underline{\beta}\cdot\underline{r} = constant$. Our solution is therefore a plane wave. The direction of the vector $\underline{\beta}$ is called direction of propagation.

Similarly, the equi-amplitude surfaces are planes normal to $\underline{\alpha}$. In particular, a plane wave is uniform when $\underline{\alpha}$ and $\underline{\beta}$ are parallel, or when $\underline{\alpha} = 0$ (whereas it can never be $\underline{\beta} = 0$, as we will see shortly).

The phase velocity in the direction of $\underline{\beta}$ is given by:

$$v_\beta = \frac{\omega}{\beta}\,.$$

while, in a given direction $\underline{r}_o$, which forms an angle $\theta < \dfrac{\pi}{2}$ with $\underline{\beta}$, the phase velocity is:

$$v_r = \frac{\omega}{\beta_r} = \frac{\omega}{\beta\cos\theta} = \frac{v_\beta}{\cos\theta} \geqslant v_\beta\,.$$

2.3 General Properties of Plane Waves

Let's consider again the separability condition:

$$\underline{k}\cdot\underline{k} = k^2 = \omega^2\mu\varepsilon_c = \omega^2\mu\left(\varepsilon - j\frac{\sigma}{\omega}\right) = \omega^2\mu\varepsilon - j\omega\mu\sigma\,,$$

in the case in which the medium is non-dispersive, so ε, μ are real and positive and σ is real and non-negative. Therefore k^2 is a complex number with positive real part and negative (or zero) imaginary part: i.e. k^2 is located in the fourth quadrant (or it lies on the real positive semi-axis) on the complex plane. k, the square root of k^2, will have two (opposite) possible positions, one being in the fourth quadrant (or lying on the positive real semi-axis) and the other being in the second quadrant (or lying on the negative real semi-axis). By convention the first determination is chosen, so we can put:

$$k = k_R - jk_J \qquad \text{con} \ \ k_R > 0 \ \ \text{and} \ \ k_J \geqslant 0 .$$

Recalling now that $\underline{k} = \underline{\beta} - j\underline{\alpha}$ we obtain:

$$(\underline{\beta} - j\underline{\alpha}) \cdot (\underline{\beta} - j\underline{\alpha}) = \beta^2 - \alpha^2 - 2j\underline{\beta} \cdot \underline{\alpha} = \omega^2 \mu \varepsilon - j\omega\mu\sigma ,$$

being $\beta^2 = \underline{\beta} \cdot \underline{\beta}$ and $\alpha^2 = \underline{\alpha} \cdot \underline{\alpha}$ as usual. Equating real parts and imaginary parts we have:

$$\begin{cases} \beta^2 - \alpha^2 = \omega^2 \mu \varepsilon \\ \underline{\beta} \cdot \underline{\alpha} = \dfrac{\omega\mu\sigma}{2} \end{cases} .$$

$\beta \neq 0$ and $\beta > \alpha$ result from the first equation. Moreover, the angle formed by $\underline{\beta}$ and $\underline{\alpha}$ is acute, being their scalar product positive.

Let us now consider the particular case of a non-dissipative medium ($\sigma = 0$). In this case it is $\underline{\beta} \cdot \underline{\alpha} = 0$. This condition can be verified either when $\underline{\alpha} = 0$, or when $\underline{\alpha}$ is orthogonal to $\underline{\beta}$. Therefore, even in the absence of losses it may exist a non-null attenuation vector $\underline{\alpha}$, provided it is orthogonal to the phase vector $\underline{\beta}$. The plane wave is uniform and not attenuated in the first case ($\underline{\alpha} = 0$), and:

$$\underline{k} \equiv \underline{\beta} = \beta \, \underline{\beta_o} = \omega\sqrt{\mu\varepsilon} \, \underline{\beta_o} = k \, \underline{\beta_o} ,$$

being $\underline{\beta_o}$ the unit vector in the direction of $\underline{\beta}$. Thus, in this particular case k represents the modulus of the propagation vector $\underline{k}$, however this property is not true in general. Moreover, the phase velocity in the direction of $\underline{\beta}$ is in this case:

$$v_\beta = \frac{\omega}{\beta} = \frac{\omega}{k} = \frac{1}{\sqrt{\mu\varepsilon}} = v ,$$

which coincides with the speed of light in the medium.[3] In a vacuum, as we have already seen, it is $c \simeq 3 \times 10^8$ [m/s]. In directions different from the one of $\underline{\beta}$ the phase velocity is larger than v.

The other possibility for lossless media, i.e. $\underline{\beta} \perp \underline{\alpha}$, describes a plane wave which is not uniform and attenuated in the direction perpendicular to the propagation direction, and with:

---

[3] Note that in our hypothesis of non-dispersive medium the phase velocity does not depend on ω.

$$\beta = \sqrt{\omega^2 \mu \varepsilon + \alpha^2} = \omega \sqrt{\mu \varepsilon + \frac{\alpha^2}{\omega^2}} > k,$$

and so:

$$v_\beta = \frac{\omega}{\beta} = \frac{1}{\sqrt{\mu \varepsilon + \frac{\alpha^2}{\omega^2}}} < v.$$

Here, moreover, the phase velocity depends on ω even though the medium is not dispersive. In this case, the phase velocity is lower than v inside a cone around the direction of β. In general, when the phase velocity is lower than the speed of light, the wave is said $slow$. Along the directions outside the cone, instead, the phase velocity is larger than v and so the wave is called *fast*.

Let us consider now the case in which the medium is dissipative and therefore $\sigma \neq 0$. In such a situation the vector $\underline{\alpha}$ must certainly be non-zero and $\underline{\beta}$ and $\underline{\alpha}$ must not be perpendicular. However, $\underline{\beta}$ and $\underline{\alpha}$ may be parallel and therefore in this case the wave is uniform and attenuated, with $\underline{\beta} = \beta \underline{\beta}_o$ and $\underline{\alpha} = \alpha \underline{\beta}_o$. In conclusion:

$$\underline{k} = \underline{\beta} - j\underline{\alpha} = (\beta - j\alpha)\underline{\beta}_o.$$

Moreover from the general equation:

$$k^2 = \beta^2 - \alpha^2 - 2j\,\underline{\beta}\cdot\underline{\alpha}$$

it follows, in this case:

$$k^2 = \beta^2 - \alpha^2 - 2j\,\beta\alpha = (\beta - j\alpha)^2 \quad \Rightarrow \quad k = \beta - j\alpha \quad \Rightarrow \quad \underline{k} = k\,\underline{\beta}_o.$$

The condition $\underline{k} = k\underline{\beta}_o$ with k real or complex, is therefore typical of uniform waves.

At this point, after considering the propagation properties, we begin to examine the polarization properties of a single plane wave, which are related, as already mentioned, only to the vector part $\underline{E}_o$. Let's start from the following relation:

$$\underline{k}\cdot\underline{E}_o = 0.$$

Putting:

$$\underline{k} = \underline{\beta} - j\underline{\alpha},$$
$$\underline{E}_o = \underline{E}_{oR} + j\underline{E}_{oJ},$$

it is:

$$(\underline{\beta} - j\underline{\alpha})\cdot(\underline{E}_{oR} + j\underline{E}_{oJ}) = 0$$

$$\Rightarrow \quad (\underline{\beta}\cdot\underline{E}_{oR} + \underline{\alpha}\cdot\underline{E}_{oJ}) + j(\underline{\beta}\cdot\underline{E}_{oJ} - \underline{\alpha}\cdot\underline{E}_{oR}) = 0.$$

Separating the real and the imaginary parts it is obtained:

$$\begin{cases} \underline{\beta}\cdot\underline{E}_{oR} + \underline{\alpha}\cdot\underline{E}_{oJ} = 0 \\ \underline{\beta}\cdot\underline{E}_{oJ} - \underline{\alpha}\cdot\underline{E}_{oR} = 0 \end{cases}.$$

Actually, these conditions are not very meaningful, at least in the general form, but they can be very significant in specific cases.

When the electric field of a plane wave is known, then the corresponding magnetic field can be determined from the first Maxwell's equation:

$$\nabla \times \underline{E} = -j\omega\mu\,\underline{H} \quad\Rightarrow\quad -j\underline{k} \times \underline{E} = -j\omega\mu\,\underline{H}$$

$$\Rightarrow\quad \underline{H} = \frac{\underline{k} \times \underline{E}}{\omega\mu} = \frac{\underline{k} \times \underline{E}_o}{\omega\mu}\,e^{-j\underline{k}\cdot\underline{r}} = \underline{H}_o\,e^{-j\underline{k}\cdot\underline{r}},$$

having set $\underline{H}_o = \dfrac{\underline{k} \times \underline{E}_o}{\omega\mu}$. Hence the magnetic field is expressed by a wave function of the same type as the one of the electric field and so it will be a solution of the equation $\nabla^2 \underline{H} + k^2 \underline{H} = 0$.

We are going to check now that $\nabla\cdot\underline{H} = 0$, which coincides, in our hypothesis, with $\underline{k}\cdot\underline{H}_o = 0$:

$$\underline{k}\cdot\underline{H}_o = \underline{k}\cdot\frac{\underline{k} \times \underline{E}_o}{\omega\mu} = \frac{\underline{k} \times \underline{k}\cdot\underline{E}_o}{\omega\mu} = 0.$$

Now let us also verify the second Maxwell's equation $\nabla \times \underline{H} = j\omega\varepsilon_c\,\underline{E}$. It is:

$$\nabla \times \underline{H} = -j\underline{k} \times \underline{H} = \frac{-j\underline{k} \times (\underline{k} \times \underline{E}_o)}{\omega\mu}\,e^{-j\underline{k}\cdot\underline{r}} =$$

$$= -\frac{j}{\omega\mu}\left[\underline{k}\,(\underline{k}\cdot\underline{E}_o) - \underline{E}_o\,(\underline{k}\cdot\underline{k})\right]e^{-j\underline{k}\cdot\underline{r}} =$$

$$= \frac{j}{\omega\mu}\,k^2\underline{E}_o\,e^{-j\underline{k}\cdot\underline{r}} = j\omega\varepsilon_c\,\underline{E}_o\,e^{-j\underline{k}\cdot\underline{r}} = j\omega\varepsilon_c\,\underline{E}.$$

When $\underline{H}$ is known, the field $\underline{E}$ is given by:

$$\nabla \times \underline{H} = j\omega\varepsilon_c\,\underline{E} \quad\Rightarrow\quad \underline{E} = \frac{1}{j\omega\varepsilon_c}(-j\underline{k}) \times \underline{H} = -\frac{1}{\omega\varepsilon_c}\underline{k} \times \underline{H}_o\,e^{-j\underline{k}\cdot\underline{r}}$$

$$\Rightarrow\quad \underline{E}_o = -\frac{\underline{k} \times \underline{H}_o}{\omega\varepsilon_c}.$$

2.3.1 Uniform Non-attenuated Plane Wave

It is studied now the particular situation of a non-dispersive and non-dissipative medium (ε and μ real positive and $\sigma = 0$), then either $\underline{\alpha} = 0$ or $\underline{\beta} \perp \underline{\alpha}$ must occur. In the first case $\underline{k} \equiv \underline{\beta}$ and the general system of equation previously seen reduces to:

$$\underline{\beta} \cdot \underline{E}_{oR} = 0 \quad , \quad \underline{\beta} \cdot \underline{E}_{oJ} = 0 .$$

Using the duality for $\underline{H}$:

$$\underline{\beta} \cdot \underline{H}_{oR} = 0 \quad , \quad \underline{\beta} \cdot \underline{H}_{oJ} = 0 .$$

In addition, from $\underline{H} = \dfrac{\underline{k} \times \underline{E}}{\omega \mu}$ it follows, being $\underline{k}$ real:

$$\underline{H}_{oR} = \frac{\underline{\beta} \times \underline{E}_{oR}}{\omega \mu} \quad , \quad \underline{H}_{oJ} = \frac{\underline{\beta} \times \underline{E}_{oJ}}{\omega \mu} ,$$

and the dual ones

$$\underline{E}_{oR} = -\frac{\underline{\beta} \times \underline{H}_{oR}}{\omega \varepsilon} \quad , \quad \underline{E}_{oJ} = -\frac{\underline{\beta} \times \underline{H}_{oJ}}{\omega \varepsilon} .$$

Some definitions widely used for electromagnetic waves are now introduced. A wave is said *transverse electric* (TE, also known as H) with respect to a certain direction, if the electric field is orthogonal to that direction. Instead, a wave is called *transverse magnetic* (TM, also known as E) with respect to a direction when the magnetic field is orthogonal to that direction. Finally, a wave is said TEM when both the electric and the magnetic fields are orthogonal with respect to that direction. For what seen before, then, the non-attenuated uniform plane wave is TEM with respect to the propagation direction.

From the above formulas, we notice that we have two groups of relations that contain either just the vectors $\underline{E}_{oR}$ and $\underline{H}_{oR}$ or the vectors $\underline{E}_{oJ}$ and $\underline{H}_{oJ}$ only, i.e., in essence, the wave can be decomposed into two independent linearly polarized waves. The wave resulting from the superposition of the two is generally elliptically polarized.

Let's consider one of the two linearly polarized waves presented, for example the first, and let's assume:

$$\underline{E}_{oR} = E_{oR} \, \underline{e}_o \quad , \quad \underline{H}_{oR} = H_{oR} \, \underline{h}_o .$$

From $\underline{\beta} \cdot \underline{E}_{oR} = 0$ it follows:

$$\underline{\beta}_o \cdot \underline{e}_o = 0 ,$$

so that $\underline{\beta}_o \times \underline{e}_o$ is a unit vector (unitary modulus), and, moreover:

$$H_{oR} \, \underline{h}_o = \frac{\beta \, E_{oR}}{\omega \mu} \underline{\beta}_o \times \underline{e}_o \, ,$$

from which we get:

$$\underline{h}_o = \underline{\beta}_o \times \underline{e}_o \, ,$$

so the three unit vectors $\underline{e}_o, \underline{h}_o, \underline{\beta}_o$ form in this order a right tri-rectangular triad, and:

$$H_{oR} = \frac{\beta \, E_{oR}}{\omega \mu} = \frac{\omega \sqrt{\mu \varepsilon}}{\omega \mu} E_{oR} = \sqrt{\frac{\varepsilon}{\mu}} \, E_{oR} \, ,$$

$$E_{oR} = \sqrt{\frac{\mu}{\varepsilon}} \, H_{oR} \, .$$

As you can see the relationship between the amplitudes of the electric field and the magnetic field is a quantity (having physical dimensions of an impedance) that depends only on the characteristics of the medium. The quantity:

$$\zeta = \sqrt{\frac{\mu}{\varepsilon}} \, ,$$

is called characteristic impedance of the medium and it is measured in *ohm* $[\Omega]$. Its particular value in a vacuum is:

$$\zeta_o = \sqrt{\frac{\mu_o}{\varepsilon_o}} \simeq 120 \, \pi \, [\Omega] \simeq 377 \, [\Omega] \, .$$

It is clearly $\omega \mu = k \zeta$, and thus:

$$\underline{H} = \frac{1}{\zeta} \, \underline{\beta}_o \times \underline{E} \, , \qquad (2.1)$$

$$\underline{E} = -\frac{k \times \underline{H}}{\omega \varepsilon} = -\frac{k}{\omega \varepsilon} \, \underline{\beta}_o \times \underline{H} = \frac{k}{\omega \varepsilon} \, \underline{H} \times \underline{\beta}_o \, .$$

Now let us observe that:

$$\frac{k}{\omega \varepsilon} = \frac{\omega \sqrt{\mu \varepsilon}}{\omega \varepsilon} = \sqrt{\frac{\mu}{\varepsilon}} = \zeta$$

$$\Downarrow$$

$$\underline{E} = \zeta \, \underline{H} \times \underline{\beta}_o \, . \qquad (2.2)$$

Note that (2.2) could also be obtained from (2.1) multiplying by $\underline{\beta}_o$ to the right and recalling that since $\underline{E}$ is transversal with respect to $\underline{\beta}_o$, it results

$$\underline{\beta}_o \times \underline{E} \times \underline{\beta}_o = \underline{E} \, .$$

Or we could proceed by multiplying to the left (2.1) by $\underline{\beta}_o$, having:

$$\underline{\beta}_o \times \underline{H} = \frac{1}{\zeta} \underline{\beta}_o \times \left(\underline{\beta}_o \times \underline{E} \right) = -\frac{1}{\zeta} \underline{E} \, ,$$

i.e.

$$\underline{E} = \zeta \underline{H} \times \underline{\beta}_o \, .$$

In fact in general it is (note that in this case the brackets at the first member are required)

$$\underline{v}_o \times \left(\underline{v}_o \times \underline{A} \right) = \underline{v}_o \left(\underline{v}_o \cdot \underline{A} \right) - \underline{A} \left(\underline{v}_o \cdot \underline{v}_o \right) = - \left(\underline{A} - A_v \, \underline{v}_o \right) = -\underline{A}_\perp$$

(being $\underline{A}_\perp$ the orthogonal component of $\underline{A}$ with respect to the direction of $\underline{v}_o$).

Finally, let us consider the expression of the Poynting vector for this kind of wave:

$$\underline{P} = \frac{1}{2} \underline{E} \times \underline{H}^* = \frac{1}{2} \underline{E}_o \, e^{-j\underline{\beta}\cdot\underline{r}} \times \underline{H}_o^* \, e^{j\underline{\beta}\cdot\underline{r}} = \frac{1}{2} \underline{E}_o \times \frac{\underline{\beta} \times \underline{E}_o^*}{\omega\mu} =$$

$$= \frac{1}{2\omega\mu} [\underline{\beta} \, (\underline{E}_o \cdot \underline{E}_o^*) - \underline{E}_o^* \, (\underline{E}_o \cdot \underline{\beta})] = \frac{1}{2\omega\mu} \underline{\beta} \, |\underline{E}_o|^2 = \frac{1}{2 \zeta} |\underline{E}_o|^2 \underline{\beta}_o \, .$$

Therefore $\underline{P}$ is a real and constant vector directed along $\underline{\beta}$.

Similar considerations apply to the other wave associated with $\underline{E}_{oJ}$ and $\underline{H}_{oJ}$.

2.3.2 Non-uniform Plane Wave Attenuated Perpendicularly to the Direction of Propagation

Let us consider now the case of the other wave seen, the one with $\underline{\beta} \perp \underline{\alpha}$. Let us suppose first that the electric field is linearly polarized; $\underline{E}_{oR}$ and $\underline{E}_{oJ}$ are parallel in this case, and therefore they are represented by the same unit vector, which can be taken as the unit vector of the complex vector, which is multiplied by a complex amplitude.[4] It is:

$$\underline{E}_o = (E_{oR} + j \, E_{oJ}) \, \underline{e}_o \, .$$

---

[4] In fact, you can verify that this property characterizes the linearly polarized vectors.

Let us now come back to the general condition $\underline{k} \cdot \underline{E}_o = 0$ that becomes:

$$\left(\underline{\beta} - j\,\underline{\alpha}\right) \cdot (E_{oR} + j\,E_{oJ})\,\underline{e}_o = 0\,.$$

We suppress $E_{oR} + j\,E_{oJ}$, which is certainly non-zero, and then separate the real and the imaginary part obtaining

$$\underline{\beta} \cdot \underline{e}_o = 0 \quad , \quad \underline{\alpha} \cdot \underline{e}_o = 0\,.$$

Therefore $\underline{e}_o$, $\underline{\alpha}$, $\underline{\beta}$ form, in order, a right tri-rectangular triad. Moreover the relation $\underline{\beta} \cdot \underline{e}_o = 0$ indicates that the wave is TE with respect to the direction of propagation.

As regards the magnetic field, by applying the general relation for plane waves:

$$\underline{H}_o = \frac{1}{\omega\mu}\left(\underline{\beta} - j\,\underline{\alpha}\right) \times (E_{oR} + j\,E_{oJ})\,\underline{e}_o =$$

$$= \frac{E_{oR} + j\,E_{oJ}}{\omega\mu}\,\underline{\beta} \times \underline{e}_o - j\,\frac{E_{oR} + j\,E_{oJ}}{\omega\mu}\,\underline{\alpha} \times \underline{e}_o = \underline{H}_{o\alpha} + \underline{H}_{o\beta}\,,$$

being:

$$\underline{H}_{o\alpha} = \frac{E_{oR} + j\,E_{oJ}}{\omega\mu}\,\underline{\beta} \times \underline{e}_o = \frac{E_{oR} + j\,E_{oJ}}{\omega\mu}\,\beta\,\underline{\alpha}_o\,,$$

and

$$\underline{H}_{o\beta} = -j\,\frac{E_{oR} + j\,E_{oJ}}{\omega\mu}\,\underline{\alpha} \times \underline{e}_o = j\,\frac{E_{oR} + j\,E_{oJ}}{\omega\mu}\,\alpha\,\underline{\beta}_o\,.$$

In general, therefore, the vector $\underline{H}_o$ is elliptically polarized in the plane defined by $\underline{\alpha}$ and $\underline{\beta}$, then the wave is not TM with respect to the direction of propagation. The Poynting vector for this TE wave is:

$$\underline{P} = \frac{1}{2}\,\underline{E}_o\,e^{-j\,\underline{k}\cdot\underline{r}} \times \underline{H}_o^*\,e^{j\,\underline{k}^*\cdot\underline{r}} = \frac{1}{2}\,\underline{E}_o \times \frac{\underline{k}^* \times \underline{E}_o^*}{\omega\mu}\,e^{-j\,(\underline{k}-\underline{k}^*)\cdot\underline{r}} =$$

$$= \frac{1}{2\omega\mu}\left[\underline{k}^*\left(\underline{E}_o \cdot \underline{E}_o^*\right) - \underline{E}_o^*\left(\underline{E}_o \cdot \underline{k}^*\right)\right] e^{-j\,(\underline{k}-\underline{k}^*)\cdot\underline{r}}$$

$$= \frac{1}{2\omega\mu}\,\left|\underline{E}_o\right|^2\,e^{-2\underline{\alpha}\cdot\underline{r}}\,(\underline{\beta} + j\,\underline{\alpha})\,,$$

since, being $\underline{e}_o \cdot \underline{k} = 0$ it follows that $\underline{e}_o \cdot \underline{k}^* = 0$, from which $\underline{E}_o \cdot \underline{k}^* = 0$. In this case, therefore, the Poynting vector is complex, its real component is directed along $\underline{\beta}$ and its imaginary one is directed along $\underline{\alpha}$. Moreover, $\underline{P}$ is no longer constant, but it decays exponentially in the direction of $\underline{\alpha}$.

We can now for duality assume that the magnetic field is linearly polarized:

$$\underline{H}_o = (H_{oR} + j\, H_{oJ})\, \underline{h}_o ,$$

and from the general relation $\underline{k}\cdot\underline{H}_o = 0$ it follows that $\underline{\beta}\cdot\underline{h}_o = 0$ and $\underline{\alpha}\cdot\underline{h}_o = 0$. This time $\underline{h}_o$, $\underline{\alpha}$, $\underline{\beta}$ form a right tri-rectangular triad and the wave is TM with respect to the direction of propagation.

For the electric field, we have:

$$\underline{E}_o = -\frac{1}{\omega\varepsilon}\left(\underline{\beta} - j\,\underline{\alpha}\right) \times (H_{oR} + j\, H_{oJ})\, \underline{h}_o =$$

$$= -\frac{H_{oR} + j\, H_{oJ}}{\omega\varepsilon}\, \underline{\beta} \times \underline{h}_o + j\,\frac{H_{oR} + j\, H_{oJ}}{\omega\varepsilon}\, \underline{\alpha} \times \underline{h}_o = \underline{E}_{o\alpha} + \underline{E}_{o\beta} ,$$

being

$$\underline{E}_{o\alpha} = -\frac{H_{oR} + j\, H_{oJ}}{\omega\varepsilon}\, \underline{\beta} \times \underline{h}_o = -\frac{H_{oR} + j\, H_{oJ}}{\omega\varepsilon}\, \beta\, \underline{\alpha}_o ,$$

and

$$\underline{E}_{o\beta} = j\,\frac{H_{oR} + j\, H_{oJ}}{\omega\varepsilon}\, \underline{\alpha} \times \underline{h}_o = -j\,\frac{H_{oR} + j\, H_{oJ}}{\omega\varepsilon}\, \alpha\, \underline{\beta}_o .$$

So in general the electric field is elliptically polarized in the plane defined by $\underline{\alpha}$ and $\underline{\beta}$, this means that the wave is not TE with respect to the direction of propagation.

The Poynting vector is:

$$\underline{P} = \frac{1}{2}\underline{E}_o\, e^{-j\,\underline{k}\cdot\underline{r}} \times \underline{H}_o^*\, e^{j\,\underline{k}^*\cdot\underline{r}} =$$

$$= -\frac{1}{2\omega\varepsilon}\left(\underline{k} \times \underline{H}_o\right) \times \underline{H}_o^*\, e^{-j(\underline{k}-\underline{k}^*)\cdot\underline{r}}$$

$$= \frac{1}{2\omega\varepsilon}\left[\underline{k}\left(\underline{H}_o^*\cdot\underline{H}_o\right) - \underline{H}_o\left(\underline{H}_o^*\cdot\underline{k}\right)\right]\, e^{-j(\underline{k}-\underline{k}^*)\cdot\underline{r}} =$$

$$= \frac{|\underline{H}_o|^2}{2\omega\varepsilon}\, e^{-2\underline{\alpha}\cdot\underline{r}}\, \underline{k} = \frac{|\underline{H}_o|^2}{2\omega\varepsilon}\, e^{-2\underline{\alpha}\cdot\underline{r}}\left(\underline{\beta} - j\,\underline{\alpha}\right).$$

Also in this case, therefore, the Poynting vector has a real part directed along $\underline{\beta}$ and an imaginary part directed along $\underline{\alpha}$; moreover, $\underline{P}$ decays exponentially in the direction of $\underline{\alpha}$.

2.3.3 Uniform Attenuated Plane Waves

Finally, we are going to consider the case of a plane wave which is uniform, but this time attenuated, thus the medium is dissipative. In this case:

$$\underline{k} = (\beta - j\alpha)\,\underline{\beta}_o = k\underline{\beta}_o = \omega\sqrt{\mu\varepsilon_c}\,\underline{\beta}_o \,.$$

We have already seen that the condition $\underline{k} = k\underline{\beta}_o$ is the one that characterizes uniform plane waves.

From the general relation $\underline{k}\cdot\underline{E}_o = 0$ it follows that $\underline{\beta}_o\cdot\underline{E}_o = 0$. In addition, it is:

$$\underline{H}_o = \frac{k}{\omega\mu}\,\underline{\beta}_o \times \underline{E}_o \quad\Rightarrow\quad \underline{H}_o\cdot\underline{\beta}_o = 0,$$

therefore, even if the medium is dissipative, the uniform plane wave is TEM with respect to the propagation direction.

It should be observed that assuming the electric field linearly polarized, i.e. $\underline{E}_o = E_o\,\underline{e}_o$ where E_o is a scalar quantity in general complex, it is, from the general condition, $\underline{\beta}_o\cdot\underline{e}_o = 0$ and it is, for the magnetic field:

$$\underline{H}_o = \frac{kE_o}{\omega\mu}\,\underline{\beta}_o \times \underline{e}_o = H_o\underline{h}_o \,,$$

where $\underline{h}_o = \underline{\beta}_o \times \underline{e}_o$ is a unit vector, being $\underline{\beta}_o$ orthogonal to $\underline{e}_o$, while:

$$H_o = \frac{k}{\omega\mu}E_o = \frac{\omega\sqrt{\mu\varepsilon_c}}{\omega\mu}E_o = \sqrt{\frac{\varepsilon_c}{\mu}}\,E_o$$

$$\Downarrow$$

$$E_o = \sqrt{\frac{\mu}{\varepsilon_c}}\,H_o = \zeta\,H_o \,.$$

Hence, from the hypothesis of linear polarisation of the vector field $\underline{E}$, it follows that the magnetic field must be linearly polarized, too. In addition, the three unit vectors $\underline{e}_o, \underline{h}_o, \underline{\beta}_o$ form, in the order, a right tri-rectangular triad.

The quantity $\zeta = \sqrt{\frac{\mu}{\varepsilon_c}}$ represents the most general (complex) expression of the characteristic impedance of the (dissipative) medium. The following expressions are still valid for the general case of ζ complex:

$$\underline{H}_o = \frac{1}{\zeta}\,\underline{\beta}_o \times \underline{E}_o \quad, \quad \underline{E}_o = \zeta\,\underline{H}_o \times \underline{\beta}_o \,.$$

The Poynting vector is expressed by the formula:

$$\underline{P} = \frac{1}{2}\,E_o\,\underline{e}_o\,e^{-j\beta\underline{\beta}_o\cdot\underline{r}}\,e^{-\alpha\underline{\beta}_o\cdot\underline{r}} \times H_o^*\,\underline{h}_o\,e^{j\beta\underline{\beta}_o\cdot\underline{r}}\,e^{-\alpha\underline{\beta}_o\cdot\underline{r}} = \frac{1}{2}\,\zeta\,|H_o|^2\,e^{-2\alpha\underline{\beta}_o\cdot\underline{r}}\,\underline{\beta}_o \,.$$

It is a complex vector directed as $\underline{\beta}_o$ and which is attenuated exponentially in the same direction.

The initial hypothesis of linearly polarized electric field actually does not affect the generality of the treatment, because any uniform plane wave, elliptically polarized in general, can be expressed as the sum of two linearly polarized waves. Establishing, for example, a Cartesian system xy on an equi-phase surface, we can consider a first wave having $\underline{E}$ polarized along $\underline{x}_o$ (and then necessarily with $\underline{H}$ along $\underline{y}_o$), and a second wave with $\underline{E}$ polarized along $\underline{y}_o$ (and $\underline{H}$ necessarily polarized along $\underline{x}_o$).

2.4 Spectrum of Plane Waves

We are going to show now, explicitly, the importance of plane waves as building blocks that can be used, by superposition, to express any kind of wave (e.g. cylindrical, spherical and so on).

We will consider a simple, homogeneous, isotropic, non-dispersive and non-dissipative medium. We have already seen how the Helmholtz equation could be obtained by a Fourier transformation of the d'Alembert equation with respect to time. It follows naturally to extend the procedure to the other (spatial) coordinates, too. For example, with respect to the variable x, it is:

$$\underline{E}(k_x, y, z, \omega) = \mathscr{F}_x\left[\underline{E}(x, y, z, \omega)\right] = \int_{-\infty}^{+\infty} \underline{E}(x, y, z, \omega)\, e^{jk_x x}\, dx\,,$$

and the inverse:

$$\underline{E}(x, y, z, \omega) = \frac{1}{2\pi} \int_{-\infty}^{+\infty} \underline{E}(k_x, y, z, \omega)\, e^{-jk_x x}\, dk_x\,.$$

The Helmholtz equation is written in Cartesian coordinates:

$$\frac{\partial^2 \underline{E}}{\partial x^2} + \frac{\partial^2 \underline{E}}{\partial y^2} + \frac{\partial^2 \underline{E}}{\partial z^2} + k^2 \underline{E} = 0$$

$$\Downarrow$$

$$-k_x^2\, \underline{E} + \frac{\partial^2 \underline{E}}{\partial y^2} + \frac{\partial^2 \underline{E}}{\partial z^2} + k^2 \underline{E} = 0\,.$$

The second derivative with respect to x is transformed, as it is well known, in a multiplication by $-k_x^2$. We need to apply the double inverse transformation with respect to k_x and ω in order to get back the field in the time domain.

At this point we transform with respect to y, too, and the differential equation becomes ordinary:

$$\frac{d^2 \underline{E}}{dz^2} + (k^2 - k_x^2 - k_y^2)\, \underline{E} = 0\,.$$

where $\underline{E}(k_x, k_y, z, \omega)$ is the unknown field from which we can recover the field in time domain through a triple anti-transformation:

$$\underline{E}(x, y, z, t) = \frac{1}{(2\pi)^3} \int_{-\infty}^{+\infty} \int_{-\infty}^{+\infty} \int_{-\infty}^{+\infty} \underline{E}(k_x, k_y, z, \omega) \, e^{j(\omega t - k_x x - k_y y)} \, dk_x \, dk_y \, d\omega \, .$$

By transforming with respect to the variable z, too, it is obtained:

$$(k^2 - k_x^2 - k_y^2 - k_z^2) \underline{E} = 0 \, ,$$

where $\underline{E}(k_x, k_y, k_z, \omega)$, four time transformed, is unknown, and from which the actual field might be obtained by

$$\underline{E}(x, y, z, t) = \frac{1}{(2\pi)^4} \int_{-\infty}^{+\infty} \int_{-\infty}^{+\infty} \int_{-\infty}^{+\infty} \int_{-\infty}^{+\infty} \underline{E}(k_x, k_y, k_z, \omega)$$
$$e^{j(\omega t - k_x x - k_y y - k_z z)} \, dk_x \, dk_y \, dk_z \, d\omega \, .$$

This is the expression of the spectrum of plane waves, which shows how a generic electric field, provided four times Fourier transformable with respect to spatial variables and time, may be expressed as a superposition (integral) of plane waves of infinitesimal amplitude:

$$\frac{1}{(2\pi)^4} \, \underline{E}(k_x, k_y, k_z, \omega) dk_x \, dk_y \, dk_z \, d\omega \, ,$$

constant, obviously, with respect to x, y, z, t.

We note, however, that such a generic electric field must not only be a function four times Fourier transformable, but it must also be a wave, i.e. a solution of the Helmholtz equation, so the quadruple Fourier transform needs to satisfy the

$$(k^2 - k_x^2 - k_y^2 - k_z^2)\underline{E} = 0 \, ,$$

which admits solutions different from the trivial one identically zero only if

$$k^2 - k_x^2 - k_y^2 - k_z^2 = 0 \Rightarrow \omega^2 \mu \varepsilon = k^2 = k_x^2 + k_y^2 + k_z^2 \, .$$

Hence we found again, in this four times transformed domain, the condition of separability which resulted, in the case of a single plane wave, from the assumption of separation of variables for the Helmholtz equation.

It would therefore be a mistake applying the Fourier inverse transform with respect to the four variables, because they are not independent. For example, we can choose not to transform with respect to the variable z. In this case we can solve the ordinary differential equation previously seen, which, in case of

$$(k^2 - k_x^2 - k_y^2) \neq 0 \, ,$$

admits the general solution

$$\underline{E}(k_x, k_y, z, \omega) = \underline{E}_1(k_x, k_y, \omega)\, e^{-j\sqrt{k^2 - k_x^2 - k_y^2} \cdot z} + \underline{E}_2(k_x, k_y, \omega)\, e^{j\sqrt{k^2 - k_x^2 - k_y^2} \cdot z} ,$$

which can be put in the triple integral seen before to obtain the field $\underline{E}(x, y, z, t)$.

We need to recall that it is not allowed to have k_x, k_y, k_z, ω all real in dissipative media, as it is required by the definition of the Fourier transform. Therefore the Laplace transform needs to be used in place of the Fourier transform.

It is also evident that not all the plane waves of the spectrum are uniform: in fact, outside the so-called circle of "visibility" $k_x^2 + k_y^2 = \omega^2 \mu \varepsilon$ the waves are not uniform and they are attenuated along $\underline{z}_o$. It is $k_z = -j\alpha_z$ and therefore

$$\underline{\beta} = k_x\, \underline{x}_o + k_y\, \underline{y}_o \quad , \quad \underline{\alpha} = \alpha_z\, \underline{z}_o \quad , \quad \underline{\beta} \perp \underline{\alpha}.$$

However, for large values of z the non-uniform (non-homogeneous, or evanescent, as it is called in the literature) portion is attenuated to negligible levels.

The plane-wave spectrum considered is also said angular spectrum, because the angle of the elementary plane wave of the spectrum changes when the wave numbers change.

2.4.1 Electric Field of a Monochromatic Wave in a Half-Space

Let us suppose that we want to find the expression for the electric field of a monochromatic wave in a half-space ($z \geq 0$). In this case the triple integral, seen for the general case, becomes a double integral with respect to a pair of wave numbers, for example k_x and k_y. Moreover, since the region of interest is infinite, only the elementary plane waves traveling in the positive z direction can be considered for $z \geq 0$. For the phasor of the electric field it is:

$$\underline{E}(x, y, z) = \frac{1}{(2\pi)^2} \int_{-\infty}^{\infty} \int_{-\infty}^{\infty} \underline{E}(k_x, k_y)\, e^{-j\,(k_x\, x + k_y\, y)}\, e^{-jk_z\, z}\, dk_x\, dk_y ,$$

with $k_z = \sqrt{\omega^2 \mu \varepsilon - k_x^2 - k_y^2}$. On the plane $z = 0$:

$$\underline{E}(x, y, 0) = \frac{1}{(2\pi)^2} \int_{-\infty}^{\infty} \int_{-\infty}^{\infty} \underline{E}(k_x, k_y)\, e^{-j\,(k_x\, x + k_y\, y)}\, dk_x\, dk_y ,$$

i.e. the electric field on the plane $z = 0$ is the double inverse Fourier transform of the spectrum amplitude function $\underline{E}(k_x, k_y)$. Reversing the transformation we have:

$$\underline{E}(k_x, k_y) = \mathscr{F}\left[\underline{E}(x, y, 0)\right] = \int_{-\infty}^{\infty} \int_{-\infty}^{\infty} \underline{E}(x, y, 0)\, e^{j\,(k_x\, x + k_y\, y)}\, dx\, dy .$$

Essentially then, to obtain the electric field in a half-space it is sufficient to know the field itself on the plane $z = 0$ (compare to the Huygens principle for the elementary spherical waves, which as the plane waves constitute a so-called complete set which can express any field). We therefore have:

$$\underline{E}(x, y, z) = \frac{1}{(2\pi)^2} \int\int_{-\infty}^{\infty} \left[\int\int_{-\infty}^{\infty} \underline{E}(x', y', 0)\, e^{j(k_x x' + k_y y')} dx'\, dy' \right]$$
$$e^{-j(k_x x + k_y y)} e^{-jk_z z} dk_x\, dk_y .$$

Rearranging the previous expression:

$$\underline{E}(x, y, z) = \frac{1}{(2\pi)^2} \int\int_{-\infty}^{\infty} \underline{E}(x', y', 0)$$
$$\left\{ \int\int_{-\infty}^{\infty} e^{-j\,[k_x\,(x-x') + k_y\,(y-y') + k_z\,z]}\, dk_x\, dk_y \right\} dx'\, dy' .$$

The quantity in brackets is said *direct propagator* $h^z(x - x', y - y')$. It then results:

$$\underline{E}(x, y, z) = \frac{1}{(2\pi)^2} \int_{-\infty}^{\infty}\int_{-\infty}^{\infty} h^z(x - x', y - y')\, \underline{E}(x', y', 0)\, dx'\, dy',$$

i.e. the propagated field is expressed by a double convolution between the field on the plane $z = 0$ and the h^z propagator.

So the process of propagation from the plane $z = 0$ up to a generic plane $z = constant > 0$ has, on the field $\underline{E}(x', y', 0)$, the same effect produced by the passage through a linear stationary system characterized by an impulse response h^z. From the impulse response in time domain we can derive a transfer function $H^z(k_x, k_y)$. In order to determine its expression we write $\underline{E}(x, y, z)$ too as a double inverse Fourier transform, in the form:

$$\underline{E}(x, y, z) = \frac{1}{(2\pi)^2} \int_{-\infty}^{\infty}\int_{-\infty}^{\infty} \left[\underline{E}(k_x, k_y)\, e^{-jk_z z} \right] e^{-j\,(k_x x + k_y y)}\, dk_x\, dk_y =$$
$$= \frac{1}{(2\pi)^2} \int_{-\infty}^{\infty}\int_{-\infty}^{\infty} \underline{E}^z(k_x, k_y)\, e^{-j\,(k_x x + k_y y)}\, dk_x\, dk_y ,$$

having written $\underline{E}^z(k_x, k_y) = \underline{E}(k_x, k_y)\, e^{-jk_z z}$. This choice is equivalent to a reference-plan translation from $z = 0$ to $z = constant > 0$. The transfer function is therefore:

$$H^z(k_x, k_y) = e^{-jk_z z} .$$

2.5 Non-monochromatic Plane Waves

We previously introduced the concept of phase velocity for a generic wave. We then presented in detail the plane waves. In the case of a uniform plane wave in a non-dissipative medium the phase constant (modulus of the phase vector) is $\beta = \omega \sqrt{\mu \varepsilon}$ that is a linear function of frequency in the case of non-dispersive medium. The phase velocity in the direction of β coincides with the speed of light in the medium ($v_\beta = \frac{\omega}{\beta} = \frac{1}{\sqrt{\mu \varepsilon}} = v$), which in the given hypothesis is a constant.

In the general case of dispersive medium suitable diagrams, called dispersion diagrams or Brillouin diagrams (in honor of the French physicist who studied these phenomena), are usually considered. They normally show the angular frequency (or the frequency) as the abscissa, and the wave number or phase constant as the ordinate, alternatively the wave number is the abscissa and the angular frequency is the ordinate: the latter is the classic version.

It is clear that in the case of non-dispersive medium the diagram above is a straight line with a certain slope ϕ with respect to the abscissa axis, and it is clearly:

$$v_\beta = \frac{\omega}{\beta} = \tan \phi.$$

In the case of dispersive medium, instead, the diagram is a curve and the angle ϕ in the previous relation (and hence the phase velocity) depends on ω. As already mentioned, the consequence of this relation is that the individual frequency components of a non-monochromatic electromagnetic field (like a modulated signal used in telecommunications) propagate with different phase velocities, and therefore the field configuration changes in general during propagation.

2.5.1 Beat Velocity

The previous considerations may be clarified by considering at first the beat phenomenon, i.e. looking at the field which results from the superposition of two monochromatic uniform plane waves having different frequency but equal amplitude and both propagating in the z direction. The instantaneous electric fields of this configuration are:

$$\underline{E}_1(z, t) = \mathrm{Re}[\underline{E}_o e^{j(\omega_1 t - \beta_1 z)}],$$

$$\underline{E}_2(z, t) = \mathrm{Re}[\underline{E}_o e^{j(\omega_2 t - \beta_2 z)}].$$

At this point we can put:

$$\overline{\omega} = \frac{\omega_1 + \omega_2}{2},$$

$$\Delta \omega = \omega_2 - \omega_1.$$

It follows:

$$\omega_1 = \frac{1}{2}(\omega_1 + \omega_2) - \frac{1}{2}(\omega_2 - \omega_1) = \overline{\omega} - \frac{\Delta\omega}{2},$$

$$\omega_2 = \frac{1}{2}(\omega_1 + \omega_2) + \frac{1}{2}(\omega_2 - \omega_1) = \overline{\omega} + \frac{\Delta\omega}{2}.$$

Then we can assume:

$$\overline{\beta} = \frac{\beta_1 + \beta_2}{2},$$

$$\Delta\beta = \beta_2 - \beta_1.$$

$$\beta_1 = \overline{\beta} - \frac{\Delta\beta}{2},$$

$$\beta_2 = \overline{\beta} + \frac{\Delta\beta}{2}.$$

Observe that, in general, $\overline{\beta} \neq \beta(\overline{\omega})$, because the dispersion law is not generally linear.

For the total field, we have:

$$\underline{E}(z, t) = \underline{E}_1(z, t) + \underline{E}_2(z, t)$$

$$= \text{Re}\{\underline{E}_o\, e^{j[(\overline{\omega} - \frac{\Delta\omega}{2})t - (\overline{\beta} - \frac{\Delta\beta}{2})z]} + \underline{E}_o\, e^{j[(\overline{\omega} + \frac{\Delta\omega}{2})t - (\overline{\beta} + \frac{\Delta\beta}{2})z]}\} =$$

$$= \text{Re}\{\underline{E}_o\, e^{j(\overline{\omega}t - \overline{\beta}z)}\, [e^{-j(\frac{\Delta\omega}{2}t - \frac{\Delta\beta}{2}z)} + e^{j(\frac{\Delta\omega}{2}t - \frac{\Delta\beta}{2}z)}]\}$$

$$= 2\cos\left(\frac{\Delta\omega}{2}t - \frac{\Delta\beta}{2}z\right)\,\text{Re}[\underline{E}_o\, e^{j(\overline{\omega}t - \overline{\beta}z)}].$$

The resulting field can then be seen as a plane wave having angular frequency equal to the average of the two angular frequencies and phase constant equal to the average of the two phase constants. Moreover, the field amplitude (real) is not constant, but it is modulated by a $2\cos(\frac{\Delta\omega}{2}t - \frac{\Delta\beta}{2}z)$ factor. The described effect is named beat.

The phase velocity of this wave in the z direction is $\overline{\omega}/\overline{\beta}$. The amplitude of the wave constitutes the envelope of the beat and varies in space and time, but it moves in the z direction in a rigid shape fashion. We can define the envelope speed as the speed of a hypothetical observer who moves in the z direction in such a way that he does not observe amplitude variations. Hence, for this observer it must be:

$$\frac{\Delta\omega}{2}t - \frac{\Delta\beta}{2}z = constant$$

$$\Downarrow$$

$$\Delta\omega\, dt - \Delta\beta\, dz = 0$$

$$\Downarrow$$

$$\Delta \omega \, dt = \Delta \beta \, dz$$

$$\Downarrow$$

$$\frac{dz}{dt} = \frac{\Delta \omega}{\Delta \beta} = v_b \,.$$

The quantity v_b is named beat velocity.

2.5.2 Group Velocity of a Wave Packet

Let us consider now the more general case of a field depending on the variables z and t represented, as inverse Fourier transform, by a spectrum of uniform plane waves propagating in the z direction:

$$\underline{E}(z, t) = \frac{1}{2\pi} \int_{-\infty}^{+\infty} \underline{E}_o(\omega) \, e^{-j\beta(\omega) z} e^{j\omega t} \, d\omega$$

(This is a particular case of the spectra previously seen in the absence of the wave numbers k_x and k_y; we chose to integrate with respect to ω, but of course we could have integrated with respect to β).

Let us make the further assumptions that the amplitude $\underline{E}_o(\omega)$ is different from zero only in the range $\omega_1 \leq \omega \leq \omega_2$ and that the field is a so called "wave packet", i.e. the following condition holds:

$$\omega_2 - \omega_1 \ll \omega_o \quad \text{with} \quad \omega_o = \frac{\omega_1 + \omega_2}{2} \,.$$

Let us now expand the (dispersion) function $\beta = \beta(\omega)$ as a Taylor's series of initial point ω_o truncated to the first-order term:

$$\beta(\omega) \simeq \beta(\omega_o) + \frac{d\beta}{d\omega}\bigg|_{\omega_o} (\omega - \omega_o) = \beta_o + \frac{d\beta}{d\omega}\bigg|_{\omega_o} \Delta \omega \,,$$

having assumed

$$\beta_o = \beta(\omega_o) \,, \quad \text{different from} \quad \frac{\beta(\omega_1) + \beta(\omega_2)}{2}$$

$$\Delta \omega = \omega - \omega_o \Rightarrow \omega = \omega_o + \Delta \omega \,.$$

Introducing the result into the integral above it is obtained:

$$\underline{E}(z,t) = \frac{1}{2\pi} \int_{\omega_1}^{\omega_2} \underline{E}_o(\omega)\, e^{-j(\beta_o + \frac{d\beta}{d\omega}\big|_{\omega_o}\Delta\omega)z}\, e^{j(\omega_o + \Delta\omega)t}\, d\omega =$$

$$= \left[\frac{1}{2\pi} \int_{\omega_1}^{\omega_2} \underline{E}_o(\omega) e^{j\Delta\omega(t - \frac{d\beta}{d\omega}\big|_{\omega_o}z)}\, d\omega\right] e^{-j\beta_o z} e^{j\omega_o t} .$$

This expression can once again be seen as a uniform plane wave having angular frequency equal to the center value of the band, phase constant corresponding to such angular frequency and phase velocity equal to ω_o/β_o, but having non-constant (and complex) amplitude. The complex amplitude factor represents the complex modulation envelope.

We can define again, as for the case of the beat, a velocity of the envelope that moves as in a rigid shape fashion, i.e. the speed at which a hypothetical observer must move in the direction of z in order to observe no variations in the (complex) amplitude. It must therefore be (varying z and t),

$$t - \frac{d\beta}{d\omega}\bigg|_{\omega_o} z = constant \Rightarrow dt - \frac{d\beta}{d\omega}\bigg|_{\omega_o} dz = 0 \Rightarrow$$

$$\Rightarrow dt = \frac{d\beta}{d\omega}\bigg|_{\omega_o} dz \Rightarrow \frac{dz}{dt} = \frac{1}{\frac{d\beta}{d\omega}\big|_{\omega_o}} = v_g .$$

This speed is called *group velocity* of the wave packet. If the function $\beta = \beta(\omega)$ is invertible, $\omega = \omega(\beta)$ being the inverse function, it is $v_g = \frac{d\omega}{d\beta}\big|_{\beta_o}$ (Fig. 2.1).

Note that in a non-dispersive medium the speed v_g is independent of ω_o and coincides with the phase velocity in the z direction. In fact, when all the component waves propagate at the same speed, their envelope propagates at that speed, too.

Fig. 2.1 Group velocity

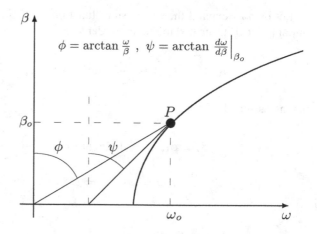

Note that if the band is not narrow, the Taylor expansion of the dispersion function can no longer be truncated to the first order, the packet is therefore deformed during propagation. Other speeds need to be introduced in this case (theory of Brillouin's precursors).

When the propagation phenomenon is no longer unidirectional, i.e. when the direction of propagation of the waves constituting the packet varies, it is necessary to generalize the treatment. Let us recall the spectrum of plane waves represented by a triple integral, but this time we choose not to integrate with respect to the angular frequency, but with respect to the three wave numbers. It is obtained:

$$\underline{E}(x, y, z, t) = \frac{1}{(2\pi)^3} \int_{-\infty}^{+\infty} \int_{-\infty}^{+\infty} \int_{-\infty}^{+\infty} \underline{E}_o(\underline{k}) \, e^{j[\omega(\underline{k})t - \underline{k}\cdot\underline{r}]} \, dk_x \, dk_y \, dk_z .$$

Now let us suppose that the amplitude $\underline{E}_o(\underline{k}) = \underline{E}_o(k_x, k_y, k_z)$ results different from zero only for

$$k_{x1} \leq k_x \leq k_{x2} \quad , \quad k_{y1} \leq k_y \leq k_{y2} \quad , \quad k_{z1} \leq k_z \leq k_{z2},$$

and let us assume, moreover:

$$k_{x2} - k_{x1} \ll |k_{xo}| ,$$

being:

$$k_{xo} = \frac{k_{x1} + k_{x2}}{2} ;$$

it is assumed, similarly:

$$k_{y2} - k_{y1} \ll |k_{yo}| \quad , \quad k_{z2} - k_{z1} \ll |k_{zo}| ,$$

with

$$k_{yo} = \frac{k_{y1} + k_{y2}}{2} \quad , \quad k_{zo} = \frac{k_{z1} + k_{z2}}{2} .$$

The described field is a plane-wave packet, and so the function of the three variables $\omega(\underline{k}) = \omega(k_x, k_y, k_z)$ can be expressed as a Taylor series of initial point $\underline{k}_o = \underline{x}_o k_{xo} + \underline{y}_o k_{yo} + \underline{z}_o k_{zo}$ truncated at the first order:

$$\omega(\underline{k}) \simeq \omega(\underline{k}_o) + \frac{\partial\omega}{\partial k_x}\bigg|_{\underline{k}_o} (k_x - k_{xo}) + \frac{\partial\omega}{\partial k_y}\bigg|_{\underline{k}_o} (k_y - k_{yo}) + \frac{\partial\omega}{\partial k_z}\bigg|_{\underline{k}_o} (k_z - k_{zo}) .$$

Defining now the vector $\underline{\Delta k} = \underline{k} - \underline{k}_o = \underline{x}_o(k_x - k_{xo}) + \underline{y}_o(k_y - k_{yo}) + \underline{z}_o(k_z - k_{zo})$ $\Rightarrow \underline{k} = \underline{k}_o + \underline{\Delta k}$ we obtain:

$$\omega(\underline{k}) \simeq \omega(\underline{k}_o) + \frac{\partial\omega}{\partial k_x}\bigg|_{\underline{k}_o} \underline{x}_o\cdot\underline{\Delta k} + \frac{\partial\omega}{\partial k_y}\bigg|_{\underline{k}_o} \underline{y}_o\cdot\underline{\Delta k} + \frac{\partial\omega}{\partial k_z}\bigg|_{\underline{k}_o} \underline{z}_o\cdot\underline{\Delta k} .$$

Let us also define the operator (nabla in the space of wave numbers):

$$\nabla_k = \underline{x}_o \frac{\partial}{\partial k_x} + \underline{y}_o \frac{\partial}{\partial k_y} + \underline{z}_o \frac{\partial}{\partial k_z},$$

obtaining the compact notation:

$$\omega(\underline{k}) \simeq \omega(\underline{k}_o) + \nabla_k \omega \Big|_{\underline{k}_o} \cdot \Delta \underline{k}.$$

We can then rewrite the exponential that appears in the spectral integral, using the above notation, as:

$$e^{j[\omega(\underline{k})\, t - \underline{k}\cdot\underline{r}]} \simeq e^{j[\omega(\underline{k}_o)\, t - \underline{k}_o \cdot \underline{r}]} \, e^{j(\nabla_k \omega |_{\underline{k}_o} \cdot \Delta \underline{k}\, t - \Delta \underline{k}\cdot\underline{r})}$$

$$= e^{j[\omega(\underline{k}_o)\, t - \underline{k}_o\cdot\underline{r}]} \, e^{j(\nabla_k \omega |_{\underline{k}_o}\, t - \underline{r})\cdot\Delta \underline{k}}.$$

Finally, the field in the time domain is:

$$\underline{E}(x, y, z, t) = \left[\frac{1}{(2\pi)^3} \int_{k_{x1}}^{k_{x2}} \int_{k_{y1}}^{k_{y2}} \int_{k_{z1}}^{k_{z2}} \underline{E}_o(\underline{k}) e^{j(\nabla_k \omega |_{\underline{k}_o}\, t - \underline{r}) \cdot \underline{\Delta k}} \, dk_x \, dk_y \, dk_z \right]$$

$$e^{j[\omega(\underline{k}_o)\, t - \underline{k}_o\cdot\underline{r}]}.$$

We obtained again a modulated uniform plane wave having wave vector $\underline{k}_o$ and angular frequency $\omega(\underline{k}_o)$. The phase velocity in a certain direction $\underline{r}_o$ is $\frac{\omega(\underline{k}_o)}{k_{0r}}$, as usual.

The envelope speed can again be defined as the speed of a hypothetical observer who does not see changes in amplitude while varying $\underline{r}$ and t. It needs to be $\nabla_k \omega \Big|_{\underline{k}_o} t - \underline{r} = \underline{constant}$, from which differentiating, it is obtained:

$$\nabla_k \omega \Big|_{\underline{k}_o} dt - d\underline{r} = 0 \Rightarrow d\underline{r} = \nabla_k \omega \Big|_{\underline{k}_o} dt \Rightarrow \frac{d\underline{r}}{dt} = \nabla_k \omega \Big|_{\underline{k}_o},$$

which represents the group velocity of the plane-wave packet just considered. Differently from the phase velocity, the group velocity is defined in general as a vector.

2.5.3 Bandwidth Relation

We are going to demonstrate now the following relation between a certain frequency range Δf and the corresponding variation of wavelength $\Delta\lambda$: $\frac{\Delta f}{f} = \frac{\Delta\lambda}{\lambda}$. We know that $\omega = 2\pi f \Rightarrow \Delta\omega = 2\pi\Delta f$. On the other hand $v = \lambda f$, and setting $\Delta f = f_2 - f_1$ assuming $f_2 > f_1$, $\Delta\lambda = \lambda_1 - \lambda_2$ with $\lambda_1 > \lambda_2$

$$\frac{\Delta f}{f} = \frac{f_2 - f_1}{f_1} = \frac{\frac{v}{\lambda_2} - \frac{v}{\lambda_1}}{\frac{v}{\lambda_1}} = \frac{\frac{\lambda_1 v - \lambda_2 v}{\lambda_1 \lambda_2}}{\frac{v}{\lambda_1}} = \frac{\lambda_1 - \lambda_2}{\lambda_2} \approx \frac{\Delta \lambda}{\lambda}.$$

The ratio $\frac{\Delta f}{f}$ is important for evaluating the performances of a resonator, defined by the so-called merit factor or quality factor Q, a parameter that is inversely proportional to $\frac{\Delta f}{f}$, where Δf is, in this case, the conventional width of the resonance peak (for example the 3-decibels width from the maximum or from the minimum, or the half-height width). The higher the Q, the narrower the peak and thus the resonance curve is selective. The ideal case would be the Dirac delta function, the more realistic case is the Lorentzian curve, such as the one previously seen for $\varepsilon_j(\omega)$, which in fact represents loss phenomena due to resonance absorption.

2.6 Reflection and Transmission of Plane Waves at a Plane Interface: Normal Incidence

So far we considered the propagation of plane waves in free space. Let us now review the effects of the presence of a flat surface which separates two half-spaces occupied by different media, both supposed homogeneous, isotropic, generally dispersive and dissipative. We start from the simpler case of normal incidence along the z (positive) direction. A uniform plane wave, coming from a medium 1 having constants $\varepsilon_1, \mu_1, \sigma_1$ and directed to a medium 2 having constants $\varepsilon_2, \mu_2, \sigma_2$ is considered.

We limit the study to the case of linear polarization, as the most general elliptical polarization can be expressed as a superposition of two linear polarizations. Therefore, we can consider the electric field of the incident wave polarized along the x direction (or we can rather choose the x direction as coincident with the polarization direction of the electric field).

In all the problems in presence of obstacles, called scattering (diffusion) or diffraction problems, we always refer to the incident field as the field that we would have in absence of obstacles. This field is assumed known. So in some way the incident field is an ideal field: the presence of an obstacle, which in this very simple case is an infinite flat interface, imposes additional conditions (boundary conditions) that can not be satisfied by the incident field only and that require the presence of another portion of the total field in medium 1, in addition to the incident wave: the so-called reflected wave. The field in medium 2 is called transmitted field and assumes different characteristics.

Therefore, for the incident electric field we have:

$$\underline{E}^i(x, y, z) = \underline{E}^i_o e^{-j\underline{k}^i \cdot \underline{r}},$$

with

$$\underline{E}^i_o = E^i_o \underline{e}^i_o = E^i_o \underline{x}_o,$$

$$\underline{k}^i = k_1 \underline{\beta}^i_o = k_1 \underline{z}_o = \omega \sqrt{\mu_1 \varepsilon_{c1}} \, \underline{z}_o.$$

It follows, as previously discussed, that also the incident magnetic field must be linearly polarized and so:

$$\underline{H}^i = \underline{H}_o^i \, e^{-j\underline{k}^i \cdot \underline{r}},$$

with

$$\underline{H}_o^i = H_o^i \, \underline{h}_o^i,$$

$$\underline{h}_o^i = \underline{\beta}_o^i \times \underline{e}_o^i = \underline{z}_o \times \underline{x}_o = \underline{y}_o.$$

Moreover, the impedance relation between the amplitudes holds, i.e.:

$$E_o^i = \zeta_1 H_o^i = \sqrt{\frac{\mu_1}{\varepsilon_{c1}}} \, H_o^i.$$

In the hypothesis of absence of surface currents on the $z = 0$ plane (assuming that none of the two media is perfect conductor), the boundary conditions require the continuity of the tangential components of both the electric and the magnetic fields and thus it is possible to determine the electromagnetic field of the two unknown waves (the reflected and the transmitted ones).

Let us suppose, for the sake of symmetry, that such waves are plane. We can write the reflected field as follows:

$$\underline{E}^r(x, y, z) = \underline{E}_o^r \, e^{-j\underline{k}^r \cdot \underline{r}},$$

$$\underline{H}^r(x, y, z) = \underline{H}_o^r \, e^{-j\underline{k}^r \cdot \underline{r}},$$

and the transmitted field:

$$\underline{E}^t(x, y, z) = \underline{E}_o^t \, e^{-j\underline{k}^t \cdot \underline{r}},$$

$$\underline{H}^t(x, y, z) = \underline{H}_o^t \, e^{-j\underline{k}^t \cdot \underline{r}}.$$

Note that the existence itself of the boundary conditions for $z = 0$

$$\underline{n} \times (\underline{E}_2 - \underline{E}_1) = 0,$$
$$\underline{n} \times (\underline{H}_2 - \underline{H}_1) = 0,$$

($\underline{n}$ directed from medium 1 to medium 2, i.e. $\underline{n} \equiv \underline{z}_o$), which need to be satisfied for all the points of the plane (and at any instant), implies that the law of spatial variation (and also the angular frequency, in the case of sinusoidal regime) is the same for $z = 0$. So it will be:

$$k_x^r = k_x^t = k_x^i,$$

$$k_y^r = k_y^t = k_y^i;$$

hence, from $k_x^i = k_y^i = 0$ it follows that the reflected and transmitted waves travel in z direction and that they are uniform (plane) waves (and therefore TEM with respect to z direction). We therefore have:

$$\underline{k}^r = k_1 \underline{\beta}_o^r = -k_1 \underline{z}_o \,,$$

because the reflected wave propagates in the direction of negative z. For the transmitted wave we have:

$$\underline{k}^t = k_2 \underline{\beta}_o^t = k_2 \underline{z}_o = \omega \sqrt{\mu_2 \varepsilon_{c2}} \, \underline{z}_o \,.$$

With the previous assumptions, the boundary conditions become:

$$\underline{z}_o \times [\underline{E}_o^t - (\underline{E}_o^i + \underline{E}_o^r)] = 0 \,,$$

$$\underline{z}_o \times [\underline{H}_o^t - (\underline{H}_o^i + \underline{H}_o^r)] = 0 \,.$$

As regards the polarization of the reflected and transmitted fields, being null the component along y of the electric field of the incident wave, there is no reason why the electric fields of the reflected and transmitted waves may have non-zero components along y, so:

$$\underline{E}_o^r = E_o^r \, \underline{x}_o \,,$$

$$\underline{E}_o^t = E_o^t \, \underline{x}_o \,,$$

and hence the reflected and transmitted waves are linearly polarized, too. The boundary condition for the electric field becomes:

$$\underline{z}_o \times \underline{x}_o \, [E_o^t - (E_o^i + E_o^r)] = 0$$

$$\Downarrow$$

$$E_o^t = E_o^i + E_o^r \,.$$

As for the magnetic field, the unit vectors are:

$$\underline{h}_o^r = \underline{\beta}_o^r \times \underline{e}_o^r = -\underline{z}_o \times \underline{x}_o = -\underline{y}_o \quad \Rightarrow \quad \underline{H}_o^r = -H_o^r \, \underline{y}_o \,,$$

$$\underline{h}_o^t = \underline{\beta}_o^t \times \underline{e}_o^t = \underline{z}_o \times \underline{x}_o = \underline{y}_o \quad \Rightarrow \quad \underline{H}_o^t = H_o^t \, \underline{y}_o \,.$$

The boundary condition for the magnetic field then becomes:

$$\underline{z}_o \times \underline{y}_o \, [H_o^t - (H_o^i - H_o^r)] = 0$$

$$\Downarrow$$

$$H_o^t = H_o^i - H_o^r .$$

Using the impedance relation it results:

$$\frac{E_o^t}{\zeta_2} = \frac{E_o^i}{\zeta_1} - \frac{E_o^r}{\zeta_1} .$$

We have finally obtained a linear system of two linear equations in the two unknowns E_o^r and E_o^t, which allows us to fully determine the reflected and transmitted amplitudes; it is obtained, by combining the two equations:

$$\frac{E_o^i - E_o^r}{\zeta_1} = \frac{E_o^i + E_o^r}{\zeta_2}$$

$$\Rightarrow E_o^r \left(\frac{1}{\zeta_1} + \frac{1}{\zeta_2} \right) = E_o^i \left(\frac{1}{\zeta_1} - \frac{1}{\zeta_2} \right) .$$

Defining the reflection coefficient of the electric field Γ_E as the ratio between the electric-field (complex) amplitudes of the reflected wave and the incident one, it is obtained:

$$\Gamma_E = \frac{E_o^r}{E_o^i} = \frac{\frac{1}{\zeta_1} - \frac{1}{\zeta_2}}{\frac{1}{\zeta_1} + \frac{1}{\zeta_2}} = \frac{\zeta_2 - \zeta_1}{\zeta_2 + \zeta_1} = \frac{\frac{\zeta_2}{\zeta_1} - 1}{\frac{\zeta_2}{\zeta_1} + 1}$$

$$\Downarrow$$

$$E_o^r = \Gamma_E \, E_o^i \quad \text{and} \quad E_o^t = (1 + \Gamma_E) \, E_o^i .$$

We therefore conclude that, in the case of normal incidence, reflection always occurs if the two media are different. In the case of oblique incidence, however, as we will see shortly, there is a condition in which total transmission occurs (the Brewster angle).

Similarly a transmission coefficient for the electric field can be defined:

$$T_E = \frac{E_o^t}{E_o^i} = 1 + \Gamma_E = 1 + \frac{\zeta_2 - \zeta_1}{\zeta_2 + \zeta_1} = \frac{\zeta_2 + \zeta_1 + \zeta_2 - \zeta_1}{\zeta_2 + \zeta_1} = \frac{2\zeta_2}{\zeta_2 + \zeta_1}$$

$$= \frac{2}{1 + \frac{\zeta_2}{\zeta_1}} \frac{\zeta_2}{\zeta_1} = \frac{2}{1 + \frac{\zeta_1}{\zeta_2}} .$$

and analogous coefficients are defined for the magnetic field:

$$\Gamma_H = \frac{H_o^r}{H_o^i} = \frac{\frac{E_o^r}{\zeta_1}}{\frac{E_o^i}{\zeta_1}} = \frac{E_o^r}{E_o^i} = \Gamma_E ,$$

$$T_H = \frac{H_o^t}{H_o^i} = \frac{\frac{E_o^t}{\zeta_2}}{\frac{E_o^i}{\zeta_1}} = \frac{\zeta_1}{\zeta_2} \frac{E_o^t}{E_o^i} = \frac{\zeta_1}{\zeta_2} T_E = \frac{\zeta_1}{\zeta_2} (1 + \Gamma_H) = \frac{2\zeta_1}{\zeta_2 + \zeta_1} = \frac{2}{1 + \frac{\zeta_2}{\zeta_1}} .$$

As a final remark it should be noted that the expressions obtained for the reflection and transmission coefficients are valid independently of the polarization of the incident wave. In the case of oblique incidence, however, we will see that the expressions of the coefficients will depend on the polarization.

Consider now the particular case of non-dispersive media, in which medium 1 is a non-dissipative dielectric ($\sigma_1 = 0$) and the medium 2 is a good conductor ($\sigma_2 \gg \omega\varepsilon_2$). Note that $\sigma_2 \gg \omega\varepsilon_1$, too, because the dielectric constants of common materials differ by at most one or two orders of magnitude. Finally, let us suppose that neither of the two media is ferromagnetic $\Rightarrow \mu_2 \simeq \mu_1$. It follows:

$$\zeta_1 = \sqrt{\frac{\mu_1}{\varepsilon_1}} ,$$

$$\zeta_2 = \sqrt{\frac{\mu_2}{\varepsilon_{c2}}} = \sqrt{\frac{j\omega\mu_2}{j\omega\varepsilon_{c2}}} = \sqrt{\frac{j\omega\mu_2}{\sigma_2 + j\omega\varepsilon_2}} \simeq \sqrt{\frac{j\omega\mu_1}{\sigma_2}} = \sqrt{j} \sqrt{\frac{\omega\mu_1}{\sigma_2}} .$$

Let's now remember that:

$$\sqrt{j} = j^{\frac{1}{2}} = (e^{j\frac{\pi}{2}})^{\frac{1}{2}} = e^{j\frac{\pi}{4}} = \cos\frac{\pi}{4} + j \sin\frac{\pi}{4} = \frac{\sqrt{2}}{2} + j\frac{\sqrt{2}}{2} = \frac{(1+j)}{\sqrt{2}}$$

$$\Downarrow$$

$$\zeta_2 \simeq (1+j)\sqrt{\frac{\omega\mu_1}{2\sigma_2}} .$$

So the impedance ratio can be expressed in magnitude:

$$\left|\frac{\zeta_2}{\zeta_1}\right| \simeq \left|(1+j)\sqrt{\frac{\omega\mu_1}{2\sigma_2}} \sqrt{\frac{\varepsilon_1}{\mu_1}}\right| = \sqrt{2}\sqrt{\frac{\omega\varepsilon_1}{2\sigma_2}} = \sqrt{\frac{\omega\varepsilon_1}{\sigma_2}} \ll 1 .$$

It is then valid the approximation:

$$\Gamma_E = \Gamma_H \simeq -1 ,$$

and

$$T_E \simeq 0 \quad , \quad T_H \simeq 2 .$$

However, it is important to observe that it would be wrong to extrapolate such relations for the perfect conductor ($\sigma_2 \to \infty$) as in that case the presence of surface current density invalidates the boundary conditions previously imposed for the tangential components. The case of the perfect conductor must therefore be studied separately.

Let us then consider a non-dissipative dielectric ($\sigma_1 = 0$) as medium 1 and a perfect conductor ($\sigma_2 \to \infty$) as medium 2. Consequently, the electric and magnetic fields in the region 2 are null everywhere ($T_E = T_H = 0$). The boundary conditions simply become:

$$-\underline{z}_o \times \underline{x}_o \, (E_o^i + E_o^r) = 0 \Rightarrow E_o^r = -E_o^i \text{ and } \underline{E}_o^r = -\underline{E}_o^i$$

$$\Downarrow$$

$$\Gamma_E = -1 .$$

In this case we talk about total reflection as the modulus of the reflected field is equal to the one of the incident field. This condition, for normal incidence, occurs only in the presence of perfect conductors, while for oblique incidence can also occur for two dielectric media, if the angle of incidence is suitably chosen in a certain region and medium 1 is denser than medium 2.

For the magnetic field, using the impedance relation, we have:

$$\begin{cases} H_o^r = -H_o^i \\ \Gamma_H = -1 \end{cases} \Rightarrow \underline{H}_o^r = -H_o^r \, \underline{y}_o = H_o^i \, \underline{y}_o = \underline{H}_o^i .$$

The situation is graphically represented in the Fig. 2.2.

From the boundary condition for the tangential component of the magnetic field we can now determine the extent of the surface currents:

$$\underline{J}_S = -\underline{z}_o \times \underline{y}_o \left(H_o^i - H_o^r \right) = 2 \, H_o^i \, \underline{x}_o = 2 \frac{E_o^i}{\zeta_1} \, \underline{x}_o .$$

In region 1, therefore, the total electric field and the total magnetic field are expressed by:

Fig. 2.2 Total reflection in the case of normal incidence

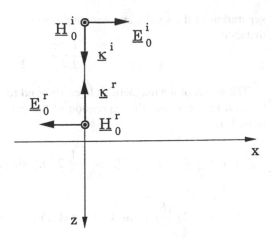

$$\underline{E}_1\,(z) = \underline{E}^i\,(z) + \underline{E}^r\,(z) = E_o^i\,\underline{x}_o\left(e^{-jk_1z} - e^{jk_1z}\right) = -2j\,E_o^i\,\sin\,(k_1z)\,\underline{x}_o\,,$$

$$\underline{H}_1\,(z) = \underline{H}^i\,(z) + \underline{H}^r\,(z) = H_o^i\,\underline{y}_o\left(e^{-jk_1z} + e^{jk_1z}\right) =$$

$$= 2\,H_o^i\,\cos\,(k_1z)\,\underline{y}_o = 2\,\frac{E_o^i}{\zeta_1}\,\cos\,(k_1z)\,\underline{y}_o\,.$$

Therefore, the total electromagnetic field represents a stationary wave, since its phase does not vary with the coordinates.

To better understand the nature of the fields, we calculate them in the time domain, too. Assuming a sinusoidal steady state, and putting $E_o^i = |E_o^i|\,e^{j\varphi}$, it is obtained:

$$\underline{E}_1\,(z,t) = \mathrm{Re}\left[\underline{E}_1\,(z)\,e^{j\omega t}\right] = \mathrm{Re}\left[-2j\,|E_o^i|\,\sin\,(k_1z)\,\underline{x}_o\,e^{j(\omega t + \varphi)}\right] =$$

$$= 2\,|E_o^i|\,\sin\,(k_1z)\,\sin\,(\omega t + \varphi)\,\underline{x}_o\,,$$

$$\underline{H}_1\,(z,t) = \mathrm{Re}\left[\underline{H}_1\,(z)\,e^{j\omega t}\right] = \mathrm{Re}\left[2\frac{|E_o^i|}{\zeta_1}\,\cos\,(k_1z)\,\underline{y}_o\,e^{j(\omega t + \varphi)}\right] =$$

$$= 2\frac{|E_o^i|}{\zeta_1}\,\cos\,(k_1z)\,\cos\,(\omega t + \varphi)\,\underline{y}_o\,.$$

The electric and magnetic fields are then orthogonal in space and in quadrature ($\frac{\pi}{2}$ phase shift) in time. This resulted also from the phasor expressions (presence of a factor $j = e^{j\frac{\pi}{2}}$). It is therefore the typical standing-wave behavior, with the separation of the spatial and temporal dependences and the presence of nodes at a distance:

$$\frac{1}{2}\frac{2\pi}{k_1} = \frac{\lambda_1}{2}.$$

The nodes of the magnetic field correspond to the anti-nodes of the electric field. Finally, let's consider the expression of the complex Poynting vector for the total field. It is:

$$\underline{P}_1(z) = \frac{1}{2}\,\underline{E}_1(z) \times \underline{H}_1^*(z) = -\frac{1}{2}\,2j\,E_o^i\,\sin(k_1z)\,\underline{x}_o \times 2\frac{E_o^{i\,*}}{\zeta_1}\cos(k_1z)\,\underline{y}_o =$$

$$= -2j\,\frac{|E_o^i|^2}{\zeta_1}\sin(k_1z)\cos(k_1z)\,\underline{z}_o = -j\,\frac{|E_o^i|^2}{\zeta_1}\sin(2k_1z)\,\underline{z}_o\,.$$

It is a purely imaginary quantity, therefore it doesn't represent a transfer of active power: the power, instead, is purely reactive. This is clearly typical of standing waves, for which actually there isn't a wave propagation.

2.7 Reflection and Transmission (Refraction) of Plane Waves at a Plane Interface: Oblique Incidence

Let us now consider the case of oblique incidence of a uniform plane wave. We immediately note that in this case the obtained results will be dependent on the polarization, which didn't occur for normal incidence. We will initially assume media at both side of the plane as isotropic, non-dispersive and non-dissipative, so that the incident uniform plane wave will be non-attenuated. We will also indicate as zx the plane of incidence, i.e. the plane containing both the normal to the separation surface and the direction of the incident plane wave. Consequently, it will be: $k_y^i = 0$ (Fig. 2.3).

It will be also:

$$k_x^i = k_1 \sin\theta^i \quad , \quad k_z^i = k_1 \cos\theta^i\,.$$

Also in this case we must admit that the incident wave excites a reflected wave in region 1 and a transmitted, or refracted, wave in region 2. We suppose, moreover, that these waves are plane.

Again, from the continuity of the tangential components of the electric and magnetic fields, the equality of the tangential wave numbers follows:

$$k_x^r = k_x^i \qquad\qquad , \qquad\qquad k_x^t = k_x^i$$
$$k_y^r = k_y^i = 0 \qquad , \qquad\qquad k_y^t = k_y^i = 0$$

Fig. 2.3 Reflection and transmission (refraction) in the case of oblique incidence

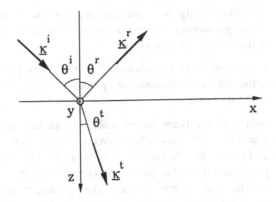

So also the reflected and transmitted wave vectors lie in the zx plane. Then:

$$k_x^r = k_1 \sin \theta^r \qquad , \qquad k_x^t = k_2 \sin \theta^t$$
$$k_z^r = -k_1 \cos \theta^r \qquad , \qquad k_z^t = k_2 \cos \theta^t$$

Note that θ^i, θ^r and θ^t angles are considered in $\left[0; \frac{\pi}{2}\right]$, so their both cosine and sine are positive.

From the relation $k_x^r = k_x^i$ it follows then:

$$k_1 \sin \theta^r = k_1 \sin \theta^i \quad \Rightarrow \quad \sin \theta^r = \sin \theta^i \quad \Rightarrow \quad \theta^r = \theta^i ,$$

which is the well-known reflection law. From $k_x^t = k_x^i$ it follows instead:

$$k_2 \sin \theta^t = k_1 \sin \theta^i , \tag{2.3}$$

which is the well-known refraction or Snell's law. The refraction angle can then be obtained:

$$\sin \theta^t = \frac{k_1}{k_2} \sin \theta^i = \frac{\omega \sqrt{\mu_1 \varepsilon_1}}{\omega \sqrt{\mu_2 \varepsilon_2}} \sin \theta^i = \frac{\mu_1}{\mu_2} \sqrt{\frac{\mu_2 \varepsilon_1}{\mu_1 \varepsilon_2}} \sin \theta^i = \frac{\mu_1}{\mu_2} \frac{\zeta_2}{\zeta_1} \sin \theta^i .$$

Note that real solutions are possible for the transmission angle only when:

$$\sqrt{\frac{\mu_1 \varepsilon_1}{\mu_2 \varepsilon_2}} \sin \theta^i \leq 1 ,$$

so initially we will have to assume that both the incidence angle and the parameters of the two media satisfy this relationship.

Summarizing, we have just seen how the "kinematic" characteristics of the so-called geometrical optics (or ray optics, which essentially involves the wave vectors and not the amplitudes), i.e.:

- the three wave vectors must lie on the same plane,
- the angles of incidence and reflection are equal,
- Snell's law,

simply derive from the existence of boundary conditions valid in all points of the $z = 0$ plane. The so-called "dynamic" properties, i.e., the amplitudes of the reflected and transmitted fields derive, instead, from the various conditions. Other dynamic properties, related to the complex amplitudes, are the phase and polarization changes in reflection and transmission. The dynamic properties are then specified by appropriate reflection and transmission coefficients which are generalizations of the similar ones considered for normal incidence; the difference is that such coefficients, named Fresnel coefficients, depend on the type of incident-wave (linear) polarization assumed.

As already seen, the incident wave, generally elliptically polarized, can always be decomposed into two linear polarizations. It will be seen that the Fresnel coefficients for the two polarizations are different. In particular, we choose now the one with the electric field directed along $\underline{y}_o$. The $\underline{E}$ field is then parallel to the separation surface and orthogonal to the plane of incidence (while the magnetic field will lie on the plane of incidence). With reference to the case of the ground, this type of polarization is called horizontal; it is also called TE with respect to z direction. The other polarization will instead have the magnetic field oriented along $\underline{y}_o$, and the electric field lying on the plane of incidence: $\underline{E}$ will then have a vertical component, and therefore this type of polarization is called vertical, it is also called parallel (to the plane of incidence) or TM with respect to z direction.

2.7.1 Horizontal Polarization

We are going to consider first the case of horizontal polarization. It is:

$$\underline{E}_o^i = E_o^i \, \underline{y}_o \,,$$

and therefore:

$$\underline{H}_o^i = \frac{1}{\omega \mu_1} \underline{k}^i \times \underline{E}_o^i = \frac{k_1}{\omega \mu_1} \left(\sin \theta^i \, \underline{x}_o + \cos \theta^i \, \underline{z}_o \right) \times E_o^i \, \underline{y}_o =$$

$$= \frac{E_o^i}{\zeta_1} \left(\sin \theta^i \, \underline{z}_o - \cos \theta^i \, \underline{x}_o \right) \,,$$

Fig. 2.4 Reflection and transmission in the case of horizontal polarization

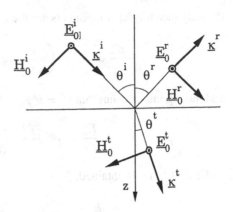

obviously, also $\underline{H}_o^i$ is linearly polarized and lies in the plane of incidence. Since the incident electric field is tangential and directed along y, for symmetry, we can also assume $\underline{E}_o^r$ and $\underline{E}_o^t$ directed along y and then $\underline{E}_o^r = E_o^r \, \underline{y}_o$, $\underline{E}_o^t = E_o^t \, \underline{y}_o$. For the respective magnetic fields, we have (Fig. 2.4):

$$\underline{H}_o^r = \frac{1}{\omega \mu_1} \underline{k}^r \times \underline{E}_o^r = \frac{k_1}{\omega \mu_1} \left(\sin \theta^r \, \underline{x}_o - \cos \theta^r \, \underline{z}_o \right) \times E_o^r \, \underline{y}_o =$$

$$= \frac{E_o^r}{\zeta_1} \left(\sin \theta^r \, \underline{z}_o + \cos \theta^r \, \underline{x}_o \right) ,$$

$$\underline{H}_o^t = \frac{1}{\omega \mu_2} \underline{k}^t \times \underline{E}_o^t = \frac{k_2}{\omega \mu_2} \left(\sin \theta^t \, \underline{x}_o + \cos \theta^t \, \underline{z}_o \right) \times E_o^t \, \underline{y}_o =$$

$$= \frac{E_o^t}{\zeta_2} \left(\sin \theta^t \, \underline{z}_o - \cos \theta^t \, \underline{x}_o \right) .$$

These two fields, too, are linearly polarized on the plane of incidence.

At this point we apply the continuity condition for the tangential electric field. This condition is identical to the one seen for normal incidence:

$$E_o^t = E_o^i + E_o^r . \tag{2.4}$$

Regarding the continuity condition for the tangential magnetic field:

$$\underline{z}_o \times \left[\underline{H}_o^t - \left(\underline{H}_o^i + \underline{H}_o^r \right) \right] = 0 ,$$

the only meaningful components are the ones along x, and here angles appear:

$$-\frac{E_o^t}{\zeta_2} \cos\theta^t - \left(-\frac{E_o^i}{\zeta_1} \cos\theta^i + \frac{E_o^r}{\zeta_1} \cos\theta^r\right) = 0,$$

and, taking into account that $\theta^r = \theta^i$,

$$\frac{E_o^t}{\zeta_2} \cos\theta^t - \frac{E_o^i - E_o^r}{\zeta_1} \cos\theta^i = 0.$$

Using (2.4) it is obtained:

$$\frac{E_o^i + E_o^r}{\zeta_2} \cos\theta^t = \frac{E_o^i - E_o^r}{\zeta_1} \cos\theta^i,$$

$$E_o^r \left(\frac{\cos\theta^t}{\zeta_2} + \frac{\cos\theta^i}{\zeta_1}\right) = E_o^i \left(\frac{\cos\theta^i}{\zeta_1} - \frac{\cos\theta^t}{\zeta_2}\right).$$

For the reflection coefficient of the electric field it is therefore:

$$\Gamma_E^h = \frac{E_o^r}{E_o^i} = \frac{\frac{\cos\theta^i}{\zeta_1} - \frac{\cos\theta^t}{\zeta_2}}{\frac{\cos\theta^i}{\zeta_1} + \frac{\cos\theta^t}{\zeta_2}} = \frac{\zeta_2 \cos\theta^i - \zeta_1 \cos\theta^t}{\zeta_2 \cos\theta^i + \zeta_1 \cos\theta^t} = \frac{\frac{\zeta_2}{\zeta_1} \cos\theta^i - \cos\theta^t}{\frac{\zeta_2}{\zeta_1} \cos\theta^i + \cos\theta^t}. \qquad (2.5)$$

For the transmission coefficient of the electric field, it is:

$$T_E^h = \frac{E_o^t}{E_o^i} = \frac{E_o^i + E_o^r}{E_o^i} = 1 + \Gamma_E^h,$$

as in the case of normal incidence. For the magnetic field, from the impedance relation it follows, in a similar way to the case of normal incidence:

$$\Gamma_H^h = \frac{H_o^r}{H_o^i} = \frac{\frac{E_o^r}{\zeta_1}}{\frac{E_o^i}{\zeta_1}} = \Gamma_E^h,$$

$$T_H^h = \frac{H_o^t}{H_o^i} = \frac{\frac{E_o^t}{\zeta_2}}{\frac{E_o^i}{\zeta_1}} = \frac{\zeta_1}{\zeta_2} \frac{E_o^t}{E_o^i} = \frac{\zeta_1}{\zeta_2} T_E^h.$$

A less general but significant expression can be obtained in the case of non-ferromagnetic media ($\mu_2 \simeq \mu_1$) for which we have:

$$\frac{\zeta_2}{\zeta_1} = \frac{\mu_2}{\mu_1} \frac{\sin\theta^t}{\sin\theta^i} \simeq \frac{\sin\theta^t}{\sin\theta^i},$$

and then (2.5) becomes[5]:

$$\Gamma_E^h \simeq \frac{\frac{\sin \theta^t}{\sin \theta^i} \cos \theta^i - \cos \theta^t}{\frac{\sin \theta^t}{\sin \theta^i} \cos \theta^i + \cos \theta^t} = \frac{\sin \theta^t \cos \theta^i - \cos \theta^t \sin \theta^i}{\sin \theta^t \cos \theta^i + \cos \theta^t \sin \theta^i} = \frac{\sin \left(\theta^t - \theta^i \right)}{\sin \left(\theta^t + \theta^i \right)},$$

which is never zero for $\theta^i \leq \frac{\pi}{2}$ (apart from the trivial case $\theta^t = \theta^i$ that occurs only if the two media are actually the same medium. So reflection always occurs, as was for the case of normal incidence. We will see instead that this is not the case for the other polarization.

2.7.2 Vertical Polarization

Let us consider now the (dual) case of vertical polarization, the situation is shown in Fig. 2.5.

For the electric field of the incident wave as a function of the magnetic field $\underline{H}_o^i = H_o^i \, \underline{y}_o$ it is:

$$\underline{E}_o^i = -\frac{1}{\omega \varepsilon_1} \underline{k}^i \times \underline{H}_o^i = -\frac{k_1}{\omega \varepsilon_1} \left(\sin \theta^i \, \underline{x}_o + \cos \theta^i \, \underline{z}_o \right) \times H_o^i \, \underline{y}_o =$$

$$= -\zeta_1 \, H_o^i \left(\sin \theta^i \, \underline{z}_o - \cos \theta^i \, \underline{x}_o \right).$$

For the same reasons seen in the case of horizontal polarization, we assume:

$$\underline{H}_o^r = H_o^r \, \underline{y}_o \quad , \quad \underline{H}_o^t = H_o^t \, \underline{y}_o \, .$$

We derive the corresponding electric fields:

$$\underline{E}_o^r = -\frac{1}{\omega \varepsilon_1} \underline{k}^r \times \underline{H}_o^r = -\frac{k_1}{\omega \varepsilon_1} \left(\sin \theta^r \, \underline{x}_o - \cos \theta^r \, \underline{z}_o \right) \times H_o^r \, \underline{y}_o =$$

$$= -\zeta_1 \, H_o^r \left(\sin \theta^r \, \underline{z}_o + \cos \theta^r \, \underline{x}_o \right),$$

$$\underline{E}_o^t = -\frac{1}{\omega \varepsilon_2} \underline{k}^t \times \underline{H}_o^t = -\frac{k_2}{\omega \varepsilon_2} \left(\sin \theta^t \, \underline{x}_o + \cos \theta^t \, \underline{z}_o \right) \times H_o^t \, \underline{y}_o =$$

$$= -\zeta_2 \, H_o^t \left(\sin \theta^t \, \underline{z}_o - \cos \theta^t \, \underline{x}_o \right).$$

---

[5] Recalling that:

$$\sin \left(\alpha \pm \beta \right) = \sin \alpha \cos \beta \pm \cos \alpha \sin \beta \, .$$

Fig. 2.5 Reflection and transmission in the case of vertical polarization

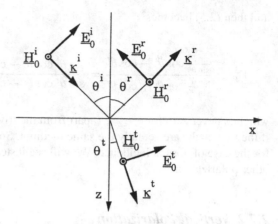

At this point we impose the continuity condition for the magnetic field, which is simply:

$$\underline{z}_o \times \left[\underline{H}_o^t - \left(\underline{H}_o^i + \underline{H}_o^r \right) \right] = 0 \quad \Rightarrow \quad H_o^t = H_o^i + H_o^r. \tag{2.6}$$

The condition for the electric field is instead:

$$\underline{z}_o \times \left[\underline{E}_o^t - \left(\underline{E}_o^i + \underline{E}_o^r \right) \right] = 0;$$

considering that only the components along $\underline{x}_o$ give contribution this becomes:

$$\zeta_2 \, H_o^t \cos \theta^t - \zeta_1 \, H_o^i \cos \theta^i + \zeta_1 \, H_o^r \cos \theta^r = 0,$$

and taking into account that $\theta^r = \theta^i$:

$$\zeta_2 \, H_o^t \cos \theta^t - \zeta_1 \cos \theta^i \left(H_o^i - H_o^r \right) = 0. \tag{2.7}$$

Combining (2.6) and (2.7) together we obtain:

$$\zeta_2 \left(H_o^i + H_o^r \right) \cos \theta^t = \zeta_1 \cos \theta^i \left(H_o^i - H_o^r \right),$$

$$H_o^r \left(\zeta_2 \cos \theta^t + \zeta_1 \cos \theta^i \right) = H_o^i \left(\zeta_1 \cos \theta^i - \zeta_2 \cos \theta^t \right),$$

$$\Gamma_H^v = \frac{H_o^r}{H_o^i} = \frac{\zeta_1 \cos \theta^i - \zeta_2 \cos \theta^t}{\zeta_2 \cos \theta^t + \zeta_1 \cos \theta^i} = \frac{\cos \theta^i - \frac{\zeta_2}{\zeta_1} \cos \theta^t}{\cos \theta^i + \frac{\zeta_2}{\zeta_1} \cos \theta^t}.$$

The transmission coefficient of the magnetic field T_H^v is obtained, taking (2.6) into account:

$$T_H^v = \frac{H_o^t}{H_o^i} = 1 + \Gamma_H^v .$$

As regards the electric-field coefficients, it is:

$$\Gamma_E^v = \frac{E_o^r}{E_o^i} = \frac{\zeta_1 H_o^r}{\zeta_1 H_o^i} = \Gamma_H^v ,$$

$$T_E^v = \frac{E_o^t}{E_o^i} = \frac{\zeta_2 H_o^t}{\zeta_1 H_o^i} = \frac{\zeta_2}{\zeta_1} T_H^v = \frac{\zeta_2}{\zeta_1} \left(1 + \Gamma_E^v\right) .$$

In the particular case $\theta^i = 0$ and hence $\theta^r = \theta^t = 0$, it is:

$$\Gamma_H^v = \frac{1 - \frac{\zeta_2}{\zeta_1}}{1 + \frac{\zeta_2}{\zeta_1}} = \frac{\zeta_1 - \zeta_2}{\zeta_2 + \zeta_1} .$$

Note that the change of sign with respect to the formula for normal incidence simply derives because in that case we used a different convention assuming $\underline{H}_o^r = -H_o^r \, \underline{y}_o$, obviously we need in any case to establish the actual signs of these quantities.

2.7.2.1 Total Transmission, Brewster Angle

In absence of ferromagnetic materials, i.e. $\mu_2 \simeq \mu_1$ and therefore:

$$\frac{\zeta_2}{\zeta_1} \simeq \frac{\sin \theta^t}{\sin \theta^i} ,$$

the following expression is obtained for Γ_E^v and Γ_H^v:

$$\Gamma_E^v = \Gamma_H^v \simeq \frac{\cos \theta^i - \frac{\sin \theta^t}{\sin \theta^i} \cos \theta^t}{\cos \theta^i + \frac{\sin \theta^t}{\sin \theta^i} \cos \theta^t} = \frac{\sin \theta^i \cos \theta^i - \sin \theta^t \cos \theta^t}{\sin \theta^i \cos \theta^i + \sin \theta^t \cos \theta^t} =$$

$$= \frac{\sin \left(2\theta^i\right) - \sin \left(2\theta^t\right)}{\sin \left(2\theta^i\right) + \sin \left(2\theta^t\right)} .$$

Then applying the trigonometry formula:

$$\frac{\sin \alpha - \sin \beta}{\sin \alpha + \sin \beta} = \frac{\tan \left(\frac{\alpha - \beta}{2}\right)}{\tan \left(\frac{\alpha + \beta}{2}\right)} ,$$

we obtain:

$$\Gamma_E^v \simeq \frac{\tan\left(\theta^i - \theta^t\right)}{\tan\left(\theta^i + \theta^t\right)} .$$

Pay attention to the fact that now there exists a value of the incidence angle for which $\Gamma_E = 0$ (total transmission), different from the trivial case $\theta^t = \theta^i$. This happens if the denominator goes to infinity, that is:

$$\theta^i + \theta^t = \frac{\pi}{2} \quad \Rightarrow \quad \theta^t = \frac{\pi}{2} - \theta^i \quad \Rightarrow \quad \sin\theta^t = \cos\theta^i ,$$

and then:

$$\frac{\sin\theta^t}{\sin\theta^i} = \frac{\cos\theta^i}{\sin\theta^i} = \frac{1}{\tan\theta^i} \simeq \frac{\zeta_2}{\zeta_1} \simeq \sqrt{\frac{\varepsilon_1}{\varepsilon_2}} \quad \Rightarrow \quad \theta^i \simeq \arctan\sqrt{\frac{\varepsilon_2}{\varepsilon_1}} = \theta_B^i .$$

This particular incidence angle takes the name of *Brewster angle* θ_B^i; it is also called angle of (linear) polarization because when a uniform plane wave, in general elliptically polarized, impinges on the separation surface at this angle, the reflected wave results linearly horizontally polarized, because the vertical polarization is completely transmitted. This is a typical example of how, in the case of oblique incidence, the polarization characteristics of the incident wave may be radically changed in reflection and transmission.

2.7.3 Total Reflection

Let us recall now that the transmission angle resulted real (regardless of the polarization type) when:

$$\sqrt{\frac{\mu_1\varepsilon_1}{\mu_2\varepsilon_2}}\ \sin\theta^i \leq 1 . \tag{2.8}$$

Being $\sin\theta^i \leq 1$, the above condition is always satisfied when $\mu_2\varepsilon_2 \geq \mu_1\varepsilon_1$. In this case medium 2 is said denser than medium 1, and from (2.3) it follows $\theta^t < \theta^i$. On the contrary, if medium 1 is denser than medium 2, as for example it happens if the electromagnetic wave is propagating within an optical fiber and the interface is with the air outside, it is $\theta^t > \theta^i$ and it will certainly happen that for suitable values of the incidence angle θ^i the first member of (2.8) will be larger than 1. In particular, the transition occurs for $\sin\theta^t = 1$ (and then $\theta^t = \frac{\pi}{2}$) in correspondence to a certain limit angle of incidence θ_L^i. So we have:

$$\sqrt{\frac{\mu_1\varepsilon_1}{\mu_2\varepsilon_2}}\ \sin\theta_L^i = 1 \quad \Rightarrow \quad \theta_L^i = \arcsin\sqrt{\frac{\mu_2\varepsilon_2}{\mu_1\varepsilon_1}} .$$

θ^t is not real anymore when $\theta^i > \theta^i_L$. In particular, we obtain a purely imaginary value for:

$$\cos \theta^t = \sqrt{1 - \sin^2 \theta^t},$$

and the angle θ^t results complex.

In this angular region, in order to resolve this contradiction keeping θ^t real, we must assume that the reflected and transmitted plane waves can't be both uniform: in particular, the transmitted wave is assumed non-uniform. Being medium 2 lossless by hypothesis, this wave can only be evanescent, i.e. $\underline{\alpha} \perp \underline{\beta}$. Therefore the wave vector is complex in medium 2:

$$\underline{k}^t = \underline{\beta}^t - j\underline{\alpha}^t,$$

with:

$$\underline{\beta}^t \cdot \underline{\alpha}^t = 0, \quad \beta^{t2} - \alpha^{t2} = k_2^2 = \omega^2 \mu_2 \varepsilon_2.$$

From the general continuity condition (which is still valid, of course):

$$k_x^t = k_x^i = k_1 \sin \theta^i;$$

it follows:

$$\beta_x^t - j\alpha_x^t = k_1 \sin \theta^i \quad \Rightarrow \quad \alpha_x^t = 0, \quad \beta_x^t = k_1 \sin \theta^i,$$

$$k_y^t = k_y^i = 0 \quad \Rightarrow \quad \beta_y^t = \alpha_y^t = 0 \quad \Rightarrow \quad \underline{\alpha}^t = \alpha^t \underline{z}_o, \quad \underline{\beta}^t = \beta^t \underline{x}_o = k_1 \sin \theta^i \underline{x}_o.$$

It can be observed that $k_1 > k_2$ because $\mu_1 \varepsilon_1 > \mu_2 \varepsilon_2$. In the hypothesis of uniform transmitted wave, the (2.3) should hold, which can not be verified for θ^i large enough to have $k_1 \sin \theta^i > k_2$, i.e. for $\theta^i > \theta^i_L$. On the other hand, the condition $k_1 \sin \theta^i = \beta^t$ must be verified for the non-uniform wave, where:

$$\beta^t = \sqrt{k_2^2 + \alpha^{t2}} > k_2,$$

and this relation can be satisfied by:

$$\alpha^t = \sqrt{\beta^{t2} - k_2^2} = \sqrt{k_1^2 \sin^2 \theta^i - k_2^2} = \omega \sqrt{\mu_1 \varepsilon_1 \sin^2 \theta^i - \mu_2 \varepsilon_2}.$$

The situation is shown in Fig. 2.6, from which we can understand why the evanescent wave is also called surface wave in the literature, as it is confined to the separation surface. About this evanescent wave it can be written:

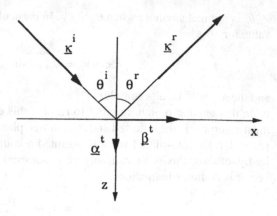

Fig. 2.6 Total reflection in the case of oblique incidence

$$\underline{E}^t\,(x,z) = \underline{E}^t_o\,e^{-j\beta^t x}\,e^{-\alpha^t z}\,,$$

$$\underline{H}^t\,(x,z) = \underline{H}^t_o\,e^{-j\beta^t x}\,e^{-\alpha^t z}\,.$$

In the case of horizontal polarization we had $\underline{E}^t_o = E^t_o\,\underline{y}_o$, and the transmitted wave was TE with respect to the propagation direction while, when the polarization is vertical, it is $\underline{H}^t_o = H^t_o\,\underline{y}_o$, and the transmitted wave is TM with respect to the propagation direction.

We could arrive at the same result by using the general formulas and accepting from the beginning that the transmission angle θ^t might be complex. It is anyway $k^t_x = k_2 \sin \theta^t$. Then, for the separability condition:

$$k^t_z = \sqrt{k^2_2 - k^{t\,2}_x} = k_2\,\sqrt{1 - \sin^2 \theta^t} = k_2\,\cos \theta^t\,.$$

It is clear, however, that the geometric interpretation of the components of a real vector loses its validity now, as they were related to real angles. In any case it is:

$$k_2\,\sin \theta^t = k_1\,\sin \theta^i\,,$$

where:

$$\theta^t = \theta^t_R + j\theta^t_J$$

is the complex angle; it follows:

$$k_1\,\sin \theta^i = k_2\,\sin \left(\theta^t_R + j\theta^t_J\right) = k_2\,\left[\sin \theta^t_R \cos \left(j\theta^t_J\right) + \cos \theta^t_R \sin \left(j\theta^t_J\right)\right]\,.$$

Fig. 2.7 Graph of the
hyperbolic sine function

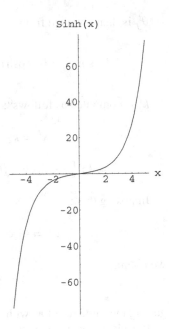

Let's now recall the following formulas:

$$\cos(jz) = \frac{e^{j(jz)} + e^{-j(jz)}}{2} = \frac{e^{-z} + e^{z}}{2} = \cosh z,$$

$$\sin(jz) = \frac{e^{j(jz)} - e^{-j(jz)}}{2j} = \frac{e^{-z} - e^{z}}{2j} = -\frac{1}{j}\sinh z = j \sinh z,$$

$$k_1 \sin\theta^i = k_2\left(\sin\theta^t_R \cosh\theta^t_J + j\cos\theta^t_R \sinh\theta^t_J\right),$$

from which, equating the real and imaginary parts, the following is obtained:

$$k_1 \sin\theta^i = k_2 \sin\theta^t_R \cosh\theta^t_J,$$

$$0 = k_2 \cos\theta^t_R \sinh\theta^t_J.$$

From the hyperbolic sine graph shown in Fig. 2.7, and having to be $\theta^t_J \neq 0$ (otherwise we have again a refraction phenomenon), it follows:

$$\cos\theta^t_R = 0 \quad \Rightarrow \quad \theta^t_R = \frac{\pi}{2} \quad \Rightarrow \quad \sin\theta^t_R = 1,$$

θ_J^t is then obtained from:

$$k_1 \sin \theta^i = k_2 \cosh \theta_J^t \quad \Rightarrow \quad \theta_J^t = \cosh^{-1} \left(\sqrt{\frac{\mu_1 \varepsilon_1}{\mu_2 \varepsilon_2}} \sin \theta^i \right).$$

k_z^t is computed as follows[6]:

$$k_z^t = k_2 \cos \theta^t = k_2 \cos \left(\theta_R^t + j\theta_J^t \right) =$$

$$= k_2 \left[\cos \theta_R^t \cos \left(j\theta_J^t \right) - \sin \theta_R^t \sin \left(j\theta_J^t \right) \right] = -jk_2 \sinh \theta_J^t .$$

Imposing then:

$$\beta^t = k_2 \cosh \theta_J^t = k_x^t \quad , \quad \alpha^t = k_2 \sinh \theta_J^t$$

we obtain:

$$\underline{k}^t = \beta^t \underline{x}_o - j\alpha^t \underline{z}_o ,$$

getting the same result as with the other method.

It is clear that since the transmission angle is complex, the reflection coefficients are in general complex, too. Let us recall the expressions:

$$\Gamma_E^h = \frac{\frac{\zeta_2}{\zeta_1} \cos \theta^i - \cos \theta^t}{\frac{\zeta_2}{\zeta_1} \cos \theta^i + \cos \theta^t} ,$$

$$\Gamma_E^v = \frac{\cos \theta^i - \frac{\zeta_2}{\zeta_1} \cos \theta^t}{\cos \theta^i + \frac{\zeta_2}{\zeta_1} \cos \theta^t} .$$

Note that, in both cases, being $\cos \theta^t$ purely imaginary, we have two complex conjugate quantities at the numerator and denominator. So the magnitude of the ratio is unitary, and therefore the magnitude of the reflected wave is equal to the one of the incident wave. When this occurs, it is said that there is *total reflection*.

---

[6] Recalling that:

$$\cos (\alpha \pm \beta) = \cos \alpha \cos \beta \mp \sin \alpha \sin \beta .$$

2.7.4 Reflection and Transmission from Conductors, Leontovich's Condition

The last case that we wish to consider concerns a uniform plane wave coming from a medium having null conductivity, which hits obliquely a conductor.

We can still suppose that the reflected wave is uniform, and the condition $\theta^r = \theta^i$ still holds. Moreover, since medium 2 is dissipative, the attenuation vector will certainly be different from zero. On the other hand, since there is no wave attenuation in medium 1, from the continuity of the tangential wave numbers it follows that the attenuation in the second medium can be directed only along the z axis, and therefore it must be $\underline{\alpha}^t = \alpha^t \underline{z}_o$. The phase vector $\underline{\beta}^t$, instead, can never be directed along z, because again from the continuity condition it follows that the component along x needs to be non-zero. Therefore, the transmitted wave cannot be uniform, and neither can be $\underline{\beta}^t \perp \underline{\alpha}^t$ since $\sigma_2 \neq 0$. The condition on the magnitudes below is still valid:

$$\beta^{t2} - \alpha^{t2} = \omega^2 \mu_2 \varepsilon_2 , \tag{2.9}$$

and it is also:

$$\beta^t \alpha^t \cos \theta^t = \frac{\omega \mu_2 \sigma_2}{2} , \tag{2.10}$$

being evidently θ^t the angle between $\underline{\beta}^t$ and the normal.

From the continuity condition for $\overline{k}_x$ it is:

$$k_1 \sin \theta^i = \beta^t \sin \theta^t . \tag{2.11}$$

From the three Eqs. (2.9)–(2.11) the three unknown quantities β^t, α^t and θ^t can be obtained. In general it is $\beta^t > \alpha^t$, so the following inequality holds:

$$\beta^t \alpha^t \cos \theta^t < \beta^t \alpha^t < \beta^{t2} ,$$

then:

$$\beta^t > \sqrt{\beta^t \alpha^t \cos \theta^t} = \sqrt{\frac{\omega \mu_2 \sigma_2}{2}} ,$$

and for the transmission angle is:

$$\sin \theta^t = \frac{k_1}{\beta^t} \sin \theta^i < \frac{\omega \sqrt{\mu_1 \varepsilon_1} \sin \theta^i}{\sqrt{\frac{\omega \mu_2 \sigma_2}{2}}} = \sqrt{\frac{2\omega \mu_1 \varepsilon_1}{\mu_2 \sigma_2}} \sin \theta^i = \sqrt{\frac{2\mu_1}{\mu_2}} \sqrt{\frac{\omega \varepsilon_1}{\sigma_2}} \sin \theta^i .$$

Let us suppose now that the second medium is a good conductor ($\sigma_2 \gg \omega\varepsilon_2$ and hence also $\sigma_2 \gg \omega\varepsilon_1$, because the orders of magnitude of ε_1 and ε_2 are comparable). In addition, if we assume (as it is often the case) $\mu_1 \simeq \mu_2$, it results:

$$\sin\theta^t \ll 1 \quad \Rightarrow \quad \theta^t \simeq 0.$$

In essence, the transmitted wave is almost uniform, and therefore practically TEM in the z direction, no matter what the incidence angle θ^i and the type of polarization are. If the incident field is generic, it can still be considered as a superposition of plane waves, so the transmitted field will be the one just seen for every component wave, and therefore also the resulting field will be practically tangential. At this point we can write the relationship between the tangential fields (which is the one for uniform plane waves):

$$\underline{E}_T^+ \simeq \zeta_2 \left(\underline{H}_T^+ \times \underline{n} \right),$$

where the unit vector $\underline{n}$ enters in medium 2 and the $+$ superscript indicates that we are just below the separation surface. However, since the considered medium is not a perfect conductor, both tangential fields are continuous and the same relation can be written from the side of the dielectric medium:

$$\underline{E}_T^- \simeq \zeta_2 \left(\underline{H}_T^- \times \underline{n} \right), \tag{2.12}$$

where we have seen (in Sect. 2.6) that for a good conductor it is:

$$\zeta_2 = \sqrt{\frac{\mu_2}{\varepsilon_{c2}}} \simeq (1+j)\sqrt{\frac{\omega\mu_2}{2\sigma_2}}.$$

Hence the relation (2.12) can be taken as an approximate boundary condition for the separation surface between a non-dissipative medium and a good conductor, and it is named *Leontovich's condition* or *Schelkunoff's condition*. This relation also applies when the separation surface is not plane, as long as its curvature is not very large. In the limit $\sigma_2 \to \infty$ (2.12) tends regularly to:

$$\underline{E}_T^- = 0,$$

that is the condition for a perfect conductor.

Finally, it should be noticed that for this quasi-uniform wave it is:

$$k_2 = \omega\sqrt{\mu_2\varepsilon_{c2}} = \omega\sqrt{\mu_2 \left(\varepsilon_2 - j\frac{\sigma_2}{\omega} \right)} = \omega\sqrt{\frac{\mu_2}{j\omega}(j\omega\varepsilon_2 + \sigma_2)} \simeq \omega\sqrt{\frac{\mu_2\sigma_2}{j\omega}} =$$

$$= \sqrt{-j\omega\mu_2\sigma_2}. \tag{2.13}$$

Remember that:

$$-j = e^{-j\frac{\pi}{2}} \quad \Rightarrow \quad \sqrt{-j} = e^{-j\frac{\pi}{4}} = \cos\frac{\pi}{4} - j\sin\frac{\pi}{4} = \frac{(1-j)}{\sqrt{2}},$$

and so (2.13) becomes (for a generic medium):

$$k = \beta - j\alpha \simeq (1-j)\sqrt{\frac{\omega\mu\sigma}{2}},$$

with:

$$\beta \cong \alpha \cong \sqrt{\frac{\omega\mu\sigma}{2}}.$$

It is seen that the attenuation α increases with frequency. For example for copper it is $\sigma \simeq 5.8 \times 10^7 \left[\Omega^{-1}/\text{m}\right]$, $\mu \simeq \mu_o = 4\pi \times 10^{-7}$ [H/m] and then:

$$\alpha \simeq 15\sqrt{f} \left[\text{m}^{-1}\right].$$

The penetration (or skin) depth δ is conventionally computed as the distance at which the modulus of the field is reduced by a factor $1/e$ ($\simeq 37\%$) with respect to the initial value at the interface. It is therefore equal to:

$$\delta = \frac{1}{\alpha} \simeq \sqrt{\frac{2}{\omega\mu\sigma}} \simeq \frac{0.07}{\sqrt{f}}$$

for copper. For example, for $f = 1\,\text{MHz}$ at the distance of 0.3 mm the wave is already attenuated to 1/100 of the initial value. Then a high-frequency electromagnetic field is very quickly attenuated within a metal (skin effect). The skin depth is just a characteristic parameter of a material at a given frequency and is useful not only for planar configurations, as far as the calculated value of δ is much smaller than the radius of curvature at all points of the surface. Furthermore, note that the assumption of infinite depth of medium 2 is not totally unrealistic, because, with such levels of attenuation, the field "visibility" can easily be smaller than the material thickness, which then appears virtually infinite to the field.

Chapter 3
Introduction to Transmission Lines

Abstract Transmission lines are introduced, emphasizing their fundamental role of simple mathematical and circuital mono-dimensional model, to describe the propagation phenomenon. The basic equations and properties are derived, and the various relevant quantities (impedance, admittance, reflection coefficient, standing-wave ratio) are examined. Finally, basic matching techniques are presented.

3.1 Transmission Line Equations, Primary and Secondary Constants, Boundary Conditions

Let us consider now a mathematical model for the study of propagation phenomena that was historically introduced for application to telegraph and telephone lines (the so-called telegrapher's and telephonist's equations) later generalized to many problems of guided propagation and to high-frequency circuits, where the geometric dimensions are larger than the wavelength. Such circuits are named distributed circuits or transmission lines. Let us start with an introductory example applying the described model to the propagation of uniform plane waves in a layered indefinite medium, in the direction orthogonal to the stratification.

Without loss of generality we can consider a linearly polarized plane wave propagating in a medium assumed initially homogeneous, isotropic, non-dispersive, but dissipative. The chosen reference system is such that the z axis coincides with the propagation direction, the x axis with the polarization direction of the electric field and consequently the y axis with the polarization direction of the magnetic field. We therefore have:

$$\underline{E} = E_x(z)\,\underline{x}_o \qquad \text{where} \qquad E_x(z) = E_o\,e^{-jkz}$$
$$\underline{H} = H_y(z)\,\underline{y}_o \qquad \qquad H_y(z) = H_o\,e^{-jkz} \;.$$

The complex Poynting vector is given by:

$$\underline{P} = \frac{1}{2}\,E_x(z)\,H_y^*(z)\,\underline{z}_o = P(z)\,\underline{z}_o.$$

© Springer International Publishing Switzerland 2015
F. Frezza, *A Primer on Electromagnetic Fields*,
DOI 10.1007/978-3-319-16574-5_3

In essence then, the field is described by the scalar functions $E_x(z)$ and $H_y(z)$. Let us now introduce the above assumptions into the homogeneous Maxwell's equations. Since:

$$\frac{\partial}{\partial x} = 0 \quad \text{and} \quad \frac{\partial}{\partial y} = 0 \quad \Rightarrow \quad \frac{\partial}{\partial z} = \frac{d}{dz},$$

the first equation becomes:

$$\left(\underline{z}_o \frac{d}{dz} \right) \times \left(E_x \, \underline{x}_o \right) = -j\omega\mu \, H_y \, \underline{y}_o,$$

$$\left(\underline{z}_o \frac{d}{dz} \right) \times \left(H_y \, \underline{y}_o \right) = j\omega\varepsilon_c \, \underline{x}_o \, E_x,$$

$$\Rightarrow \quad \begin{aligned} \frac{dE_x}{dz} &= -j\omega\mu \, H_y \\ \frac{dH_y}{dz} &= -j\omega\varepsilon_c \, E_x \end{aligned} \quad .$$

Assuming:

$$V(z) = E_x(z) \qquad Z_S = j\omega\mu$$
$$I(z) = H_y(z) \quad , \quad Y_P = j\omega\varepsilon_c \, ,$$

we obtain the two equations:

$$\begin{cases} \dfrac{dV(z)}{dz} = -Z_S \, I(z) \\[2mm] \dfrac{dI(z)}{dz} = -Y_P \, V(z) \end{cases} ,$$

and:

$$P(z) = \frac{1}{2} V(z) \, I^*(z).$$

The two quantities $V(z)$ and $I(z)$ are respectively named equivalent voltage and equivalent current (recall that the electric and magnetic field physical dimensions are, respectively, $\left[\frac{V}{m} \right]$ and $\left[\frac{A}{m} \right]$). The Z_S and Y_P quantities are called series impedance per unit length $\left[\frac{\Omega}{m} \right]$ and shunt admittance per unit length $\left[\frac{\Omega^{-1}}{m} \right]$. The obtained equations are called transmission line or telegrapher's equations as they were historically introduced while studying the voltage and current evolution in a two-wire line, which is a line formed by two parallel straight conductors immersed in a dielectric. In that case the voltages and currents represented the actual difference of potential between two conductors and current in the conductors.

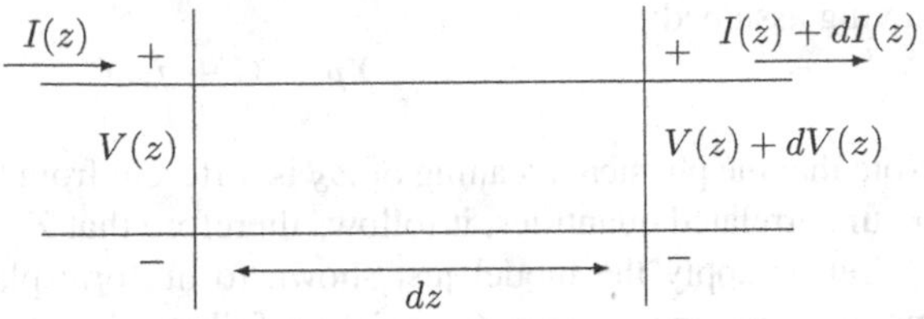

Fig. 3.1 Diagram of two-wire line

The physical circuit dimension can no longer be considered punctiform when the dimensions of the line are comparable with the wavelength (i.e., not much smaller, as it happens for low-frequency electronic circuits); in this case both the voltage and the current vary along the line (Fig. 3.1).

Note, finally, that the correspondence between voltage and current in the equivalent line and the components of the electric and magnetic field, respectively, transverse to the direction of the line is quite general.

Let us then consider two sections of a two-wire line separated by a distance dz. We assume that R, L, G, C are, respectively, the resistance (series) per unit length of the non-perfect conductors, the inductance (series) per unit length of the circuit formed by the two conductors, the conductance (shunt) per unit length due to the non-zero conductivity of the interposed dielectric, and the capacitance (shunt) per unit length between the conductors. It is obtained a distributed or non-lumped circuit, because the characteristic parameters such as resistance, inductance, conductance and capacitance are distributed along the line and they accumulate along it, so they are not concentrated in a single point (which happens instead in the case of low-frequency electronics which is modeled by lumped circuits).

The output voltage $V(z+dz) = V(z) + dV(z)$ is equal to the input voltage $V(z)$ reduced by the voltage drop in dz due to the series resistance R and inductance L. It follows:

$$V(z) + dV(z) = V(z) - (R + j\omega L)\, dz\, I(z)$$

$$\Rightarrow \quad \frac{dV(z)}{dz} = -(R + j\omega L)\, I(z) = -Z_S\, I(z),$$

having assumed:

$$Z_S = R + j\omega L.$$

The output current $I(z + dz) = I(z) + dI(z)$, similarly, is equal to the input current $I(z)$ reduced by the current drop in dz caused by the presence of the shunt conductance G and capacitance C. We therefore have:

$$I(z) + dI(z) = I(z) - (G + j\omega C)\, dz\, V(z)$$

$$\Rightarrow \quad \frac{dI(z)}{dz} = -(G + j\omega C)\, V(z) = -Y_P\, V(z),$$

having assumed:

$$Y_P = G + j\omega C.$$

Note that the physical meaning of Z_S is different from the one of Y_P, as they depend on uncorrelated quantities, it follows therefore that $Z_S \neq \frac{1}{Y_P}$.

Let us apply the model just shown to uniform plane waves in non-dispersive media; being in this case $Z_S = j\omega\mu$ it follows $R = 0$, $L = \mu$, so the resistance per unit length is zero and the inductance per unit length is μ (recall that its physical dimensions are $\left[\frac{H}{m}\right]$). Moreover we have $Y_P = j\omega\varepsilon_c = \sigma + j\omega\varepsilon$ and so $G = \sigma$ and $C = \varepsilon$ (which in fact have the physical dimensions of $\left[\frac{\Omega^{-1}}{m}\right]$ and $\left[\frac{F}{m}\right]$).

The obtained two-wire line equations are therefore identical to the ones describing the uniform plane wave propagation in unbounded space. It is then possible to name transmission line every physical system whose behavior, in respect of certain aspects, is described by the telegrapher's equations. Of course in different physical systems the quantities $V(z)$, $I(z)$, Z_S and Y_P assume different meaning and they are in general equivalent quantities. The constants Z_S and Y_P are also called *primary constants* of the line.

Note how, in some ways, the telegrapher's equations represent a simplified, one-dimensional and scalar version of the Maxwell's equations. Let's see how to solve them: the procedure has been already applied, in this case we are going to obtain the Helmholtz equation in one dimension, i.e. the equation of harmonic motion. To this aim, let us derive the first equation of the lines with respect to z; then, substituting the second:

$$\frac{d^2V}{dz^2} = -Z_S \frac{dI}{dz} = -Z_S (-Y_P V) = Z_S Y_P V.$$

Assuming:

$$Z_S Y_P = -k_z^2 \quad \Rightarrow \quad \sqrt{Z_S Y_P} = jk_z,$$

it follows:

$$\frac{d^2V}{dz^2} + k_z^2 V = 0.$$

Once solved this, the current simply follows from:

$$I(z) = -\frac{1}{Z_S}\frac{dV}{dz}.$$

The k_z constant is (apparently) called propagation constant, and it is a complex quantity in general. Let us choose the square root determination that satisfies

$$k_z = \beta_z - j\alpha_z \quad \text{with} \quad \begin{array}{l} \beta_z > 0 \quad \text{if} \quad \beta_z \neq 0 \\ \alpha_z > 0 \quad \text{if} \quad \beta_z = 0 \end{array}.$$

It is:

$$k_z^2 = -Z_S Y_P = -(R + j\omega L)(G + j\omega C) =$$

$$= -RG - j\omega RC - j\omega LG + \omega^2 LC =$$

$$= (\omega^2 LC - RG) - j\omega (LG + RC),$$

that becomes $k_z^2 = \omega^2 LC$, which is real and positive, in the absence of losses ($R = G = 0$).

Applying the formalism to our uniform plane waves, it follows:

$$k_z^2 = -j\omega\mu \, j\omega\varepsilon_c = \omega^2\mu\varepsilon_c = k^2.$$

So, the propagation constant of the mathematical model is numerically coincident with the propagation constant of the medium. We could have proceeded in a dual manner, of course, obtaining an equation of harmonic motion for the current first, and then obtaining the voltage from the current. It is well known that in the case $k_z \neq 0$ the general solution of the equation of harmonic motion can be written as traveling-wave superposition:

$$V(z) = V_o^+ e^{-jk_z z} + V_o^- e^{jk_z z} = V^+(z) + V^-(z),$$

being $V^+(z) = V_o^+ e^{-jk_z z}$ the wave traveling in the positive z direction, or direct wave, and $V^-(z) = V_o^- e^{jk_z z}$ the wave traveling in the negative z direction, or reflected wave.

Let us now compute the current:

$$I(z) = -\frac{1}{Z_S}\frac{dV}{dz} = \frac{jk_z}{Z_S}\left(V_o^+ e^{-jk_z z} - V_o^- e^{jk_z z}\right) =$$

$$= I_o^+ e^{-jk_z z} + I_o^- e^{jk_z z},$$

having assumed:

$$I_o^+ = \frac{jk_z}{Z_S} V_o^+ = \frac{V_o^+}{Z_c},$$

$$I_o^- = -\frac{jk_z}{Z_S} V_o^- = -\frac{V_o^-}{Z_c},$$

where:

$$Z_c = \frac{Z_S}{jk_z} = \frac{Z_S}{\sqrt{Z_S Y_P}} = \sqrt{\frac{Z_S}{Y_P}}.$$

Similarly:

$$I(z) = I^+(z) + I^-(z),$$

where:

$$I^+(z) = \frac{V^+(z)}{Z_c} \quad , \quad I^-(z) = -\frac{V^-(z)}{Z_c}$$

$$\Rightarrow \quad I(z) = \frac{1}{Z_c}[V^+(z) - V^-(z)].$$

Z_c constant defined above is said *characteristic impedance* of the transmission line and its physical dimensions are $[\Omega]$; its inverse $Y_c = \frac{1}{Z_c}$ $[\Omega^{-1}]$ is called *characteristic admittance*. It is, for the two-wire line:

$$Z_c = \sqrt{\frac{R + j\omega L}{G + j\omega C}},$$

which reduces to $Z_c = \sqrt{\frac{L}{C}}$ in the absence of losses. We have:

$$Z_c = \sqrt{\frac{j\omega\mu}{j\omega\varepsilon_c}} = \sqrt{\frac{\mu}{\varepsilon_c}} = \zeta,$$

in our example of application to uniform plane waves; in other words the characteristic impedance of the transmission-line model matches the characteristic impedance of the medium.

The k_z and Z_c quantities are called *secondary constants* of the line. Note that, in the absence of losses, the primary constants of the line are purely imaginary while the secondary constants are purely real in both examples of transmission lines considered. The equations of the lines can be rewritten using secondary constants in place of primary ones. The two constant pairs are related by the following:

$$Z_S = jk_z Z_c \quad \text{and} \quad jk_z Y_c = \sqrt{Z_S Y_P}\sqrt{\frac{Y_P}{Z_S}} = Y_P,$$

so the equations become:

$$\begin{cases} \dfrac{dV(z)}{dz} = -jk_z Z_c I(z) \\ \dfrac{dI(z)}{dz} = -jk_z Y_c V(z) \end{cases},$$

which are called *telephonist's equations*.

The general solution, as it is well known, can be written also in terms of standing waves. For example, we can write for the voltage:

$$V(z) = A \cos(k_z z) + B \sin(k_z z).$$

The expression for the current follows:

$$I(z) = -\frac{1}{Z_S} k_z \left[-A \sin(k_z z) + B \cos(k_z z) \right] =$$

$$= \frac{1}{jZ_c} \left[A \sin(k_z z) - B \cos(k_z z) \right].$$

The constants A and B can be expressed as a function of voltage and current in a particular section of the line, for example in the $z = 0$ section, which is usually chosen to coincide with the terminal section of the line, closed on a certain load impedance. We have essentially to impose the boundary conditions for our one-dimensional scalar model. We then have:

$$V(0) = A \qquad -jZ_c I(0) = B.$$

The same result could obviously be achieved expressing the solution in terms of traveling waves, obtaining

$$V(0) = V_o^+ + V_o^- \qquad Z_c I(0) = V_o^+ - V_o^-.$$

Adding the two equations:

$$V(0) + Z_c I(0) = 2V_o^+ \quad \Rightarrow \quad V_o^+ = \frac{1}{2} [V(0) + Z_c I(0)].$$

When the second equation is subtracted from the first, it is obtained instead:

$$V(0) - Z_c I(0) = 2V_o^- \quad \Rightarrow \quad V_o^- = \frac{1}{2} [V(0) - Z_c I(0)].$$

Let us consider the particular case of a line closed on a short circuit, i.e. V(0)=0; in our application to the propagation of uniform plane waves, this short circuit corresponds to the presence in $z = 0$ of a perfectly conducting plane transverse to the z direction. It follows then:

$$V_o^+ = \frac{1}{2} Z_c I(0) \quad , \quad V_o^- = -V_o^+,$$

$$\Rightarrow \quad V(z) = V_o^+ \left(e^{-jk_z z} - e^{jk_z z}\right) = V_o^+ \, (-2j) \, \sin(k_z \, z) =$$

$$= -j Z_c \, I(0) \, \sin(k_z \, z),$$

$$\Rightarrow \quad I(z) = \frac{1}{Z_c} \, V_o^+ \left(e^{-jk_z z} + e^{jk_z z}\right) = \frac{2V_o^+}{Z_c} \, \cos(k_z \, z) =$$

$$= I(0) \, \cos(k_z \, z).$$

Clearly $V(z)$ and $I(z)$ have the typical configuration of stationary waves in the case of a lossless line (k_z and Z_c real), moreover they are in quadrature in time (presence of a factor j, i.e. $\frac{\pi}{2}$ phase shift). The complex power of the line, which corresponds to the amplitude of the complex Poynting vector for our plane wave problem, is purely imaginary (reactive):

$$P(z) = \frac{1}{2} \, V(z) \, I^*(z) = -\frac{j}{2} \, Z_c \, I(0) \, \sin(k_z \, z) \, I^*(0) \, \cos(k_z \, z) =$$

$$= -\frac{j}{4} \, Z_c \, |I(0)|^2 \, \sin(2k_z \, z);$$

we obtained an analogous result in the case of normal incidence of uniform plane waves on a perfect conductor.

Let us consider now the dual case of an open line, i.e. such that $I(0) = 0$. We then have:

$$V_o^+ = \frac{1}{2} \, V(0) \quad , \quad V_o^- = V_o^+,$$

and so:

$$V(z) = V_o^+ \left(e^{-jk_z z} + e^{jk_z z}\right) = 2 \, V_o^+ \, \cos(k_z \, z) = V(0) \, \cos(k_z \, z),$$

$$I(z) = \frac{1}{Z_c} \, V_o^+ \left(e^{-jk_z z} - e^{jk_z z}\right) = -2j \, \frac{V_o^+}{Z_c} \, \sin(k_z \, z) = -j \frac{V(0)}{Z_c} \, \sin(k_z \, z).$$

Again, if the line is lossless, a standing-wave configuration is found.

3.2 Impedance, Admittance, Reflection Coefficient

We define, along the transmission line, two functions of the variable z that are called *impedance and admittance along the line*[1] (to be not confused with the impedance and admittance constants per unit length, and with the characteristic impedance and admittance), defined by:

---

[1] Using the one or the other depending on convenience, such as components in series for impedances, shunt components for admittances.

$$Z(z) = \frac{V(z)}{I(z)},$$

$$Y(z) = \frac{I(z)}{V(z)} = \frac{1}{Z(z)}.$$

The *voltage reflection coefficient* [2] along the line is also defined as follows:

$$\Gamma_V(z) = \frac{V^-(z)}{V^+(z)},$$

In particular, in the section $z = 0$ where the Z_L load impedance is usually put we have:

$$Z(0) = \frac{V(0)}{I(0)} = Z_L \quad , \quad \Gamma_V(0) = \frac{V_o^-}{V_o^+} = \Gamma_L.$$

The reflection coefficient is:

$$\Gamma_V(z) = \frac{V_o^- \, e^{jk_z z}}{V_o^+ \, e^{-jk_z z}} = \Gamma_V(0) \, e^{2jk_z z},$$

for a generic section z. The reader needs to pay attention to the fact that, when k_z is real (lossless line), the magnitude of the reflection coefficient keeps constant along the line:

$$|\Gamma_V(z)| = |\Gamma_V(0)| = \left| \frac{V_o^-}{V_o^+} \right| = |\Gamma_V|.$$

The $Z(z)$ and $\Gamma_V(z)$ functions are not independent of each other, it is in fact:

$$Z(z) = \frac{V(z)}{I(z)} = \frac{V^+(z) + V^-(z)}{\dfrac{1}{Z_c}[V^+(z) - V^-(z)]} =$$

$$= Z_c \frac{V^+(z) + V^-(z)}{V^+(z) - V^-(z)} = Z_c \frac{1 + \dfrac{V^-(z)}{V^+(z)}}{1 - \dfrac{V^-(z)}{V^+(z)}} =$$

$$= Z_c \frac{1 + \Gamma_V(z)}{1 - \Gamma_V(z)}.$$

---

[2] A *current reflection coefficient* Γ_I can also be defined, but the voltage one is usually employed unless the subscript "I" is specified.

The relation shown can also be reversed:

$$Z(z)\left[1 - \Gamma_V(z)\right] = Z_c\left[1 + \Gamma_V(z)\right]$$

$$\Rightarrow \quad Z(z) - Z_c = \Gamma_V(z)\left[Z(z) + Z_c\right]$$

$$\Rightarrow \quad \Gamma_V(z) = \frac{Z(z) - Z_c}{Z(z) + Z_c}.$$

In particular, for $z = 0$ (i.e. on the load):

$$Z(0) = Z_L = Z_c\frac{1 + \Gamma_V(0)}{1 - \Gamma_V(0)} = Z_c\frac{1 + \Gamma_L}{1 - \Gamma_L},$$

$$\Gamma_V(0) = \Gamma_L = \frac{Z(0) - Z_c}{Z(0) + Z_c} = \frac{Z_L - Z_c}{Z_L + Z_c}.$$

Note that $\Gamma_L = \Gamma_V(0)$ can be calculated starting from the impedance Z_L and, obviously, from the line characteristic impedance Z_c. Moreover, assuming the direct-wave amplitude V_o^+ known, the amplitude of the reflected wave V_o^- can also be found, and so $V(z)$ and $I(z)$ are fully characterized.

We now continue the examination of various types of line terminations. An interesting case is found when $Z_L = Z_c$, i.e. the load impedance coincides with the characteristic impedance. In this case, the line is said *matched*, because there are no reflections. In fact, it is $\Gamma_V(0) = 0$ and therefore $\Gamma_V(z) \equiv 0$. The line impedance is equal to the characteristic impedance in any section of the line: $Z(z) \equiv Z_c$. The matching condition is very important in many applications, a typical case is observed when the presence of a reflected wave may disturb or damage a generator; the protection of the generator is so important that often the generator itself is followed by a special component, said insulator (made by magnetized ferrite, for example) which lets the direct wave pass undisturbed, while the reflected wave is blocked. Another case of matching occurs when the line can be considered virtually infinite and uniform, because in that case no reflection occurs. Finally, a highly lossy portion of line can be considered matched, as high losses make the reflections negligible.

We have already seen that, in the case of line terminated on a short circuit, we have $Z_L = 0$, and therefore $\Gamma_L = \Gamma_V(0) = -1$ and $|\Gamma_V(z)| \equiv 1$ when k_z is real[3]; in the case of open line (open-circuit termination) it is $Z_L = \infty$, so $\Gamma_L = \Gamma_V(0) = 1$ and again $|\Gamma_V(z)| \equiv 1$ for k_z real. Whenever the condition $|\Gamma_V| = 1$ is verified, we say that we are in the presence of a *total reflection*. Besides the two cases seen, there is also a third situation of total reflection that appears when the load impedance Z_L is purely imaginary (pure reactance), i.e. $Z_L = jX_L$ (there is no dissipation of active

---

[3] Remember that in the case of normal incidence of uniform plane waves on a perfect conductor it was $\Gamma_E = -1$.

or real power, but only accumulation of reactive power takes place), and when, at the same time, the characteristic impedance Z_c is real. We have:

$$\Gamma_V(0) = \Gamma_L = \frac{jX_L - Z_c}{jX_L + Z_c} = -\frac{Z_c - jX_L}{Z_c + jX_L},$$

being $\Gamma_V(0)$ the ratio of two conjugate quantities, then $|\Gamma_V(0)| = 1$ and if k_z is real, it follows: $|\Gamma_V| \equiv 1$.

We have already introduced the formula used to obtain the reflection coefficient of a line at any section once its load is known. Now we are going to find a similar formula (less immediate) for the impedance. Starting from the relationship between reflection coefficient and impedance:

$$Z(z) = Z_c \frac{1 + \Gamma_V(z)}{1 - \Gamma_V(z)} = Z_c \frac{1 + \Gamma_V(0) e^{2jk_z z}}{1 - \Gamma_V(0) e^{2jk_z z}} =$$

$$= Z_c \frac{1 + \dfrac{Z_L - Z_c}{Z_L + Z_c} e^{2jk_z z}}{1 - \dfrac{Z_L - Z_c}{Z_L + Z_c} e^{2jk_z z}},$$

and multiplying numerator and denominator by $(Z_L + Z_c) e^{-jk_z z}$:

$$Z(z) = Z_c \frac{(Z_L + Z_c) e^{-jk_z z} + (Z_L - Z_c) e^{jk_z z}}{(Z_L + Z_c) e^{-jk_z z} - (Z_L - Z_c) e^{jk_z z}} =$$

$$= Z_c \frac{Z_L (e^{jk_z z} + e^{-jk_z z}) - Z_c (e^{jk_z z} - e^{-jk_z z})}{Z_c (e^{jk_z z} + e^{-jk_z z}) - Z_L (e^{jk_z z} - e^{-jk_z z})} =$$

$$= Z_c \frac{Z_L \cos(k_z z) - jZ_c \sin(k_z z)}{Z_c \cos(k_z z) - jZ_L \sin(k_z z)}.$$

The above formula allows, for example, to obtain the input impedance $Z_i = Z(-\ell)$ of a line section with length ℓ terminated on a given load Z_L[4] in $z = 0$ (Fig. 3.2):

$$Z_i = Z(-\ell) = Z_c \frac{Z_L \cos(k_z \ell) + jZ_c \sin(k_z \ell)}{Z_c \cos(k_z \ell) + jZ_L \sin(k_z \ell)},$$

that simply follows from the fact that cosine is an even function and sine is an odd function. In particular, assuming $Z_L = Z_c$ (matching), $Z(-\ell) = Z_c$ is obtained. In particular, assuming $Z_L = 0$ (short circuit), we have:

---

[4] The formula for the admittance is identical, as long as each impedance is replaced with the corresponding admittance.

Fig. 3.2 Line length ℓ ended
on a load Z_L

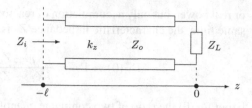

$$Z_i = jZ_c \tan(k_z \ell),$$

which is purely imaginary (reactive) if Z_c and k_z are real. In particular the reactance $X_i = Z_c \tan(k_z \ell)$ assumes alternatively positive (inductive reactance) or negative (capacitive reactance) values and it presents periodicity:

$$\frac{\pi}{k_z} = \frac{\pi}{\frac{2\pi}{\lambda_z}} = \frac{\lambda_z}{2},$$

being λ_z the wavelength of the line. In the case of open-circuit termination, or $Z_L = \infty$, it is:

$$Z_i = -jZ_c \cot(k_z \ell).$$

Finally, note that the input impedance yields the same value of the load one for any portion of line which is $\frac{\lambda_z}{2}$ long (or multiples of this value). It is in fact, for $\ell = \frac{\lambda_z}{2}$ and k_z real:

$$k_z \ell = \frac{2\pi}{\lambda_z} \frac{\lambda_z}{2} = \pi \quad \Rightarrow \quad Z_i = Z_L.$$

On the other hand, a portion of line of length $\frac{\lambda_z}{4}$ or odd multiples of this value (for even multiples obviously the previous case applies) presents an input impedance:

$$Z_i = \frac{Z_c^2}{Z_L} \quad \Rightarrow \quad Z_c = \sqrt{Z_i Z_L},$$

and the characteristic impedance results the geometric mean between the load impedance and the input impedance. This case is usually said *quarter-wave transformer*, also called *impedance inverter*: to understand why, let us introduce *normalized impedances* (dimensionless) with respect to the characteristic impedance:

$$\hat{Z}(z) = \frac{Z(z)}{Z_c} \quad \hat{Y}(z) = \frac{Y(z)}{Y_c}.$$

It is indeed:

$$\frac{Z_i}{Z_c} = \frac{Z_c}{Z_L} \quad \Rightarrow \quad \hat{Z}_i = \frac{1}{\hat{Z}_L}.$$

3.3 Standing-Wave Ratio

We now introduce another parameter, useful for the analysis of the reflection phenomena in a transmission line, which is called *standing-wave ratio* and is commonly denoted as SWR or as VSWR to clarify that it is referred to a voltage. Its definition is:

$$\text{SWR} = \frac{|V(z)|_{max}}{|V(z)|_{min}}$$

where $|V(z)|_{max}$ and $|V(z)|_{min}$ denote respectively the maximum and the minimum values of the voltage magnitude along the line. The ratio is therefore a real quantity independent of z.

Evidently, there must be a relation between this parameter and the reflection coefficient (in voltage) because the voltage magnitude is also related to such coefficient:

$$|V(z)| = |V^+(z) + V^-(z)| = |V^+(z)| |1 + \Gamma_V(z)|,$$

and, assuming k_z real:

$$|V(z)| = |V_o^+| \left|1 + \Gamma_V(0)\, e^{2jk_z z}\right| =$$

$$= |V_o^+| \left|1 + |\Gamma_V|\, e^{j\{2k_z z + \arg[\Gamma_V(0)]\}}\right|,$$

$$\text{where} \quad |\Gamma_V(0)| = |\Gamma_V|.$$

Note that the voltage magnitude is a periodic function of z with periodicity:

$$\frac{2\pi}{2k_z} = \frac{\pi}{\frac{2\pi}{\lambda_z}} = \frac{\lambda_z}{2}.$$

The maximum and minimum can therefore be found just investigating (for example with a field probe) a section of line long $\frac{\lambda_z}{2}$. We can obtain the plot drawn in the picture below representing $\frac{|V(z)|}{|V_o^+|}$ on the complex plane (Fig. 3.3).

Fig. 3.3 Representation in
the complex plane

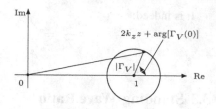

A circumference, centered in $(1, 0)$, and having $|\Gamma_V|$ radius, is described varying z. Note that if the load impedance $Z_L = R_L + jX_L$ is passive, i.e. its real part is non-negative ($R_L \geqslant 0$), and Z_o is real and positive, then we obtain:

$$|\Gamma_V| = \frac{|R_L - Z_c + jX_L|}{|R_L + Z_c + jX_L|} = \sqrt{\frac{(R_L - Z_c)^2 + X_L^2}{(R_L + Z_c)^2 + X_L^2}} \leqslant 1.$$

At this point, it follows from Fig. 3.3 inspection:

$$\frac{|V(z)|_{max}}{|V_o^+|} = 1 + |\Gamma_V|,$$

$$\frac{|V(z)|_{min}}{|V_o^+|} = 1 - |\Gamma_V|.$$

So:

$$\text{SWR} = \frac{1 + |\Gamma_V|}{1 - |\Gamma_V|}.$$

Note that it is always SWR $\geqslant 1$. In particular, SWR $= 1$ when the line is matched ($\Gamma_V = 0$, no reflection, pure traveling wave) as the voltage magnitude is in this case constant along the line. In the opposite case of total reflection, we have instead SWR $= \infty$ ($|\Gamma_V| = 1$, pure standing wave): this corresponds to the fact that in this case the magnitude of the voltage is zero in some sections of the line (nodes), and therefore $|V(z)|_{min} = 0$. The inverse relation permits to get the magnitude of the reflection coefficient starting from the SWR:

$$(1 - |\Gamma_V|)\,\text{SWR} = 1 + |\Gamma_V|,$$

$$|\Gamma_V|\,(1 + \text{SWR}) = \text{SWR} - 1 \quad \Rightarrow \quad |\Gamma_V| = \frac{\text{SWR} - 1}{\text{SWR} + 1}.$$

The SWR can be easily measured by applying its definition, i.e. just finding the maximum and the minimum magnitude of the voltage, for example, by using a probe sliding along the line, arranged in such a way that the perturbation on the field configuration is negligible.

Fig. 3.4 Anti-reflective layer (characterized by Z_{c2}, k_{z2} and ℓ) interposed between two media

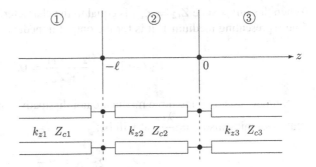

3.4 Anti-reflective Layers

It will be shown in this section how the transmission-line formalism can be used to resolve problems of transmission and reflection of uniform plane waves normally incident on layered planar structures. Every layer will be considered isotropic, non-dispersive and homogeneous, possibly lossy. An equivalent transmission line will be associated to each homogeneous layer, this line will have propagation constant, characteristic impedance and wavelength equal to the propagation constant, the characteristic impedance and the wavelength of that medium.

The tangential components of electric and magnetic fields, related to the voltage and current, respectively, must be continuous at the interface of two media, and so at the interface of two lines; it follows that voltage and current must be preserved passing through an interface as well. This means that two contiguous lines must be directly juxtaposed, without interposing any component in series or shunt (which would break the continuity of the voltage and current, respectively). From the continuity of the voltage and current the continuity of the impedance follows, and therefore the output impedance of the line on the left must coincide with the input impedance of the line on the right. The reflection coefficient, instead, is not continuous at the interface between two transmission lines.

Let us consider the so-called anti-reflective layer as a first example (Fig. 3.4). In this application a layer is interposed between two different media to suppress the reflections that would necessarily arise in correspondence to a transition section. This suppression is obtained by destructive interference between the reflected fields from the two faces of the layer.

Region 3 is supposed to be of infinite length, and then matched, it follows that its input impedance coincides with the characteristic impedance Z_{c3}, which is the output impedance for the line representing medium 2. At this point we can calculate the input impedance of line 2, which will be the load for the line 1. We have already seen the formula for this input impedance:

$$Z_{i2} = Z_{c2} \frac{Z_{c3} \cos(k_{z2}\,\ell) + j Z_{c2} \sin(k_{z2}\,\ell)}{Z_{c2} \cos(k_{z2}\,\ell) + j Z_{c3} \sin(k_{z2}\,\ell)}.$$

When this impedance $Z_{i2} = Z_{L1}$ is equal to the characteristic impedance Z_{c1} of the line representing medium 1, it is for the output impedance of line 1:

$$\Gamma_V(-\ell) = \frac{Z_{i2} - Z_{c1}}{Z_{i2} + Z_{c1}} = 0.$$

Suppose now that the three media are not dissipative, and that in particular k_{z2} is real. Assuming also $\ell = \dfrac{\lambda_{z2}}{4}$, it will be:

$$k_{z2}\,\ell = \frac{2\pi}{\lambda_{z2}} \frac{\lambda_{z2}}{4} = \frac{\pi}{2},$$

i.e. a quarter-wave transformer is implemented, for which, as we have already seen:

$$Z_{i2} = \frac{Z_{c2}{}^2}{Z_{c3}} = \frac{\zeta_2{}^2}{\zeta_3}.$$

If medium 2 is chosen so that it is:

$$\frac{\zeta_2{}^2}{\zeta_3} = Z_{c1} = \zeta_1 \quad \Rightarrow \quad \zeta_2 = \sqrt{\zeta_1\,\zeta_3}$$

then $\Gamma_V(-\ell) = 0$, i.e. there will be no reflected wave in medium 1. The anti-reflective layer 2 provides therefore the impedance matching between medium 1 and medium 3. This technique is used to achieve maximum transparency in photographic lenses. In such application medium 1 is air, medium 3 glass.

The anti-reflective layer can be dimensioned once $\lambda_{z2} = \dfrac{\lambda_o}{\sqrt{\varepsilon_{r2}}}$ is known and so ε_{r2} can be obtained (in the assumption that $\mu_r \simeq 1$ for all involved media) starting from the impedances condition:

$$\zeta_2{}^2 = \zeta_1\,\zeta_3 \quad \Rightarrow \quad \frac{\mu_o}{\varepsilon_o\,\varepsilon_{r2}} \cong \sqrt{\frac{\mu_o}{\varepsilon_o\,\varepsilon_{r1}}} \sqrt{\frac{\mu_o}{\varepsilon_o\,\varepsilon_{r3}}},$$

$$\Rightarrow \quad \frac{1}{\varepsilon_{r2}} = \frac{1}{\sqrt{\varepsilon_{r1}\,\varepsilon_{r3}}} \quad \Rightarrow \quad \varepsilon_{r2} = \sqrt{\varepsilon_{r1}\,\varepsilon_{r3}}.$$

As a second example a structure applicable when building an anechoic chamber (i.e. a chamber with non-reflective walls), or when hiding metal objects from a radar, is presented. A four-layer structure is considered in particular (Fig. 3.5).

The last layer is assumed excellent conductor (representing for example the shielding wall of the anechoic chamber or the outer wall of the radar target), hence its impedance is:

$$\zeta_4 \simeq \sqrt{\frac{j\omega\mu_4}{\sigma_4}} \simeq 0.$$

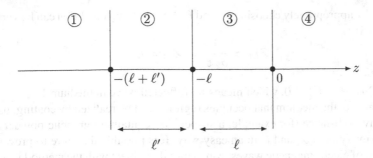

Fig. 3.5 The multilayer structure

In addition, if the thickness of region 4 is (realistically) finite, but the direct wave is sufficiently attenuated, we can assume line 4 matched and therefore line 3 terminates on $Z_{L3} = Z_{c4} = \zeta_4 \simeq 0$. So line 3 is practically short-circuited, and its input impedance is:

$$Z_{i3} = jZ_{c3}\tan(k_{z3}\,\ell).$$

Now let us suppose that medium 3 is not dissipative (k_{z3} real) and also that line 3 is a quarter-wave transformer ($\ell = \dfrac{\lambda_{z3}}{4} \Rightarrow k_{z3}\,\ell = \dfrac{\pi}{2}$). It is then $Z_{i3} = Z_{L2} = \infty$ and line 2 is terminated on an open circuit. At this point the input impedance of medium 2 is:

$$Z_{i2} = -jZ_{c2}\cot\left(k_{z2}\,\ell'\right).$$

Let us then assume that medium 2 has very small thickness (when compared, as usual, to the wavelength), so that it is:

$$|k_{z2}|\,\ell' \ll 1 \quad \Rightarrow \quad \begin{aligned} \cos\left(k_{z2}\,\ell'\right) &\simeq 1 \\ \sin\left(k_{z2}\,\ell'\right) &\simeq k_{z2}\,\ell' \end{aligned},$$

and so:

$$Z_{i2} \cong -jZ_{c2}\frac{1}{k_{z2}\,\ell'} = \frac{Z_{c2}}{jk_{z2}\,\ell'} =$$

$$= \frac{\sqrt{\dfrac{Z_{S2}}{Y_{P2}}}}{\sqrt{Z_{S2}\,Y_{P2}}\,\ell'} = \frac{1}{Y_{P2}\,\ell'} = \frac{1}{(\sigma_2 + j\omega\varepsilon_2)\,\ell'}.$$

Finally, if we assume that medium 2 is a good conductor, we get:

$$Z_{i2} = Z_{L1} \simeq \frac{1}{\sigma_2\,\ell'}.$$

Then by appropriately choosing σ_2 and ℓ' the following condition can be verified:

$$Z_{L1} \simeq \frac{1}{\sigma_2 \, \ell'} = Z_{c1} = \zeta_1,$$

and so $\Gamma_V(-\ell - \ell') = 0$, which means no reflected wave in medium 1.

In practice, the medium that occupies region 2 can be realized by coating, using a conductive substance (for example, a suspension containing graphite powder): this allows to vary both σ_2 and ℓ' in an easy way. It is possible therefore to prevent the reflection of electromagnetic waves from a metallic object with the method just seen.

It should be observed that both examples considered took advantage from the properties of quarter-wave transformer.[5] The working condition for both is said narrow-band, because when the frequency of incident radiation is changed, the wavelength also changes and therefore (for the structure already realized) $\ell = \frac{\lambda_z}{4}$ is no longer valid and the reflection coefficient is not zero. However, in practice, the reflection coefficient will be very small (almost zero) for a (narrow) band of frequencies. Many more layers would be necessary in order to obtain matching conditions for a broadband application. The theory of such structures is strongly linked to the theory of filters as they are essentially filtering structures (the signal passes for certain frequencies and it is reflected for others).

---

[5] $\ell = \frac{\lambda_z}{4}$.

Chapter 4
Guided Electromagnetic Propagation

Abstract The broad and fundamental topic of guided propagation is treated in depth, starting from general relations and introducing the important concept of guided modes (*TE*, *TM*, *TEM*). The relevant mathematical eigenvalue problem is examined, the various properties of operators, eigenvalues, eigenfunctions are reviewed, and the cut-off phenomenon is explained. The practical rectangular, circular and coaxial guides are treated. Some basic elements about cavity resonators are presented.

4.1 General Relations of Waveguides

We are going to study now particular solutions of homogeneous Maxwell's equations in structures called waveguides, which can be either dielectric or hollow metallic having cylindrical symmetry, i.e. for which the normal sections (called transverse) with respect to a given direction said axis of symmetry (also called longitudinal), are identical (both in shape and in size).

It is natural, and helpful, the use of a generalized cylindrical orthogonal coordinate system to impose the boundary conditions on this kind of structures. This coordinate system is specified with q_1, q_2 and z, in particular the Cartesian z axis is placed along the longitudinal direction, and the system of orthogonal curvilinear coordinates q_1, q_2 is established in all planes normal to z, and it will be chosen according to the shape of the transverse section. These coordinates will be, for example, the usual xy Cartesian coordinates in the case of a rectangular-section waveguide, or the polar $\rho\varphi$ coordinates for a circular cross section waveguide (such as a metal coaxial cable, or a dielectric optical fiber).

A generic vector, for example the electric field $\underline{E}$, can be expressed as follows in the chosen reference system:

$$\underline{E} = \underline{E}_t + \underline{z}_o E_z,$$

where $\underline{E}_t$ is the vector component of $\underline{E}$ in the transverse plane. In a similar way it is possible to decompose the ∇ operator and the Laplace operator in a transverse and a longitudinal component:

$$\nabla = \nabla_t + \underline{z}_o \frac{\partial}{\partial z},$$

© Springer International Publishing Switzerland 2015
F. Frezza, *A Primer on Electromagnetic Fields*,
DOI 10.1007/978-3-319-16574-5_4

$$\nabla^2 = \nabla_t^2 + \frac{\partial^2}{\partial z^2}.$$

In the particular case of Cartesian coordinates it is for example:

$$\nabla_t = \underline{x}_o \frac{\partial}{\partial x} + \underline{y}_o \frac{\partial}{\partial y},$$

$$\nabla_t^2 = \frac{\partial^2}{\partial x^2} + \frac{\partial^2}{\partial y^2}.$$

Let us consider the region of space with cylindrical symmetry in the absence of impressed sources, (i.e. in the absence of electrical and magnetic currents). The medium which occupies the region is, moreover, supposed homogeneous and isotropic. It is known that in these conditions the electric field satisfies, with the appropriate boundary conditions imposed by the structure (which we shall see later), the homogeneous Helmholtz equation:

$$\nabla^2 \underline{E} + k^2 \underline{E} = 0,$$

which, with the assumptions made, becomes:

$$\left(\nabla_t^2 + \frac{\partial^2}{\partial z^2} \right) \left(\underline{E}_t + \underline{z}_o E_z \right) + k^2 \left(\underline{E}_t + \underline{z}_o E_z \right) = 0.$$

The above equation can be modified recalling that $\underline{z}_o$ is a constant vector:

$$\nabla_t^2 \underline{E}_t + \underline{z}_o \nabla_t^2 E_z + \frac{\partial^2 \underline{E}_t}{\partial z^2} + \underline{z}_o \frac{\partial^2 E_z}{\partial z^2} + k^2 \underline{E}_t + \underline{z}_o k^2 E_z = 0.$$

Now there are only either longitudinal (parallel to $\underline{z}_o$) or transverse (orthogonal to $\underline{z}_o$) terms. By separating the two kinds of contributions it must be:

$$\begin{cases} \nabla_t^2 \underline{E}_t + \frac{\partial^2 \underline{E}_t}{\partial z^2} + k^2 \underline{E}_t = 0 \\ \nabla_t^2 E_z + \frac{\partial^2 E_z}{\partial z^2} + k^2 E_z = 0 \end{cases},$$

equivalent to the wave equation in generalized cylindrical coordinates. The second equation of the system could also be simply obtained by projecting the vectorial Helmholtz equation on the (Cartesian) z axis. The equations obtained here for the electric field can also be obtained for the magnetic field (duality). We need to write the homogeneous Maxwell's equations, too, in generalized cylindrical coordinates, in order to make them more suitable for the study of guiding structures. The first Maxwell's equation becomes therefore:

$$\left(\nabla_t + \underline{z}_o \frac{\partial}{\partial z}\right) \times \left(\underline{E}_t + \underline{z}_o E_z\right) = -j\omega\mu \left(\underline{H}_t + \underline{z}_o H_z\right).$$

$$\Rightarrow \nabla_t \times \underline{E}_t + \nabla_t \times \left(E_z \, \underline{z}_o\right) + \underline{z}_o \times \frac{\partial \underline{E}_t}{\partial z} = -j\omega\mu \, \underline{H}_t - j\omega\mu \, H_z \, \underline{z}_o. \qquad (4.1)$$

At this point we apply the following vector identity:

$$\nabla \times \left(\Phi \, \underline{A}\right) = \Phi \, \nabla \times \underline{A} - \underline{A} \times \nabla\Phi,$$

from which, in the two-dimensional case, it is:

$$\nabla_t \times \left(E_z \, \underline{z}_o\right) = E_z \, \nabla_t \times \underline{z}_o - \underline{z}_o \times \nabla_t E_z = -\underline{z}_o \times \nabla_t E_z,$$

being $\underline{z}_o$ a constant vector. From (4.1) it is then obtained:

$$\nabla_t \times \underline{E}_t - \underline{z}_o \times \nabla_t E_z + \underline{z}_o \times \frac{\partial \underline{E}_t}{\partial z} = -j\omega\mu \, \underline{H}_t - j\omega\mu \, H_z \, \underline{z}_o.$$

In order to obtain the first Maxwell's equation we split the above equation in components parallel and orthogonal to $\underline{z}_o$, obtaining:

$$\begin{cases} \nabla_t \times \underline{E}_t & = -j\omega\mu \, H_z \, \underline{z}_o \\ -\underline{z}_o \times \nabla_t E_z + \underline{z}_o \times \frac{\partial \underline{E}_t}{\partial z} & = -j\omega\mu \, \underline{H}_t \end{cases}.$$

The second Maxwell's equation is obtained by duality:

$$\begin{cases} \nabla_t \times \underline{H}_t & = j\omega\varepsilon_c \, E_z \, \underline{z}_o \\ -\underline{z}_o \times \nabla_t H_z + \underline{z}_o \times \frac{\partial \underline{H}_t}{\partial z} & = j\omega\varepsilon_c \, \underline{E}_t \end{cases}.$$

Note that the equations written so far have retained very general validity.

We are going now to consider a particular class of solutions, characterized by a separation of variables, i.e., solutions which satisfy:

$$\underline{E}_t \, (q_1, q_2, z) = \underline{e}_t \, (q_1, q_2) \, V \, (z),$$

$$\underline{H}_t \, (q_1, q_2, z) = \underline{h}_t \, (q_1, q_2) \, I \, (z),$$

in which the transversal component is characterized by the product of a vectorial factor depending only on transverse coordinates and a scalar factor depending only on the longitudinal coordinate. Evidently not all the solutions can be put into this form, but if we are able to demonstrate that these solutions form a complete set then any solution could be written in terms of them. Such a decomposition will allow us to associate an equivalent transmission line to the propagation along the longitudinal z direction.

Three classes of fields of the above type are considered: those for which $E_z \equiv 0$ (transverse electric fields, or TE); those for which $H_z \equiv 0$ (transverse magnetic fields, or TM); and those that satisfy $E_z = H_z \equiv 0$ (transverse electromagnetic fields, or TEM). These particular waves are in effect extensively used in the applications. Let us examine them separately.

4.1.1 TE Fields

In the TE case the two Maxwell's equations become:

$$\begin{cases} \nabla_t \times \underline{E}_t = -j\omega\mu\, H_z\, \underline{z}_o \\ \underline{z}_o \times \frac{\partial \underline{E}_t}{\partial z} = -j\omega\mu\, \underline{H}_t \end{cases}, \tag{4.2}$$

$$\begin{cases} \nabla_t \times \underline{H}_t = 0 \\ -\underline{z}_o \times \nabla_t H_z + \underline{z}_o \times \frac{\partial \underline{H}_t}{\partial z} = j\omega\varepsilon_c\, \underline{E}_t \end{cases}. \tag{4.3}$$

Introducing the hypothesis of variables separability in the first equation, we have:

$$V(z)\, \nabla_t \times \underline{e}_t (q_1, q_2) = -j\omega\mu\, H_z (q_1, q_2, z)\, \underline{z}_o.$$

Given the form of the first member, it can be deducted that H_z, too, can be written in separated form, i.e.:

$$H_z (q_1, q_2, z) = h_z (q_1, q_2)\, V(z).$$

So the transverse component of $\underline{E}$ and the longitudinal component of $\underline{H}$ have the same dependence on z. Simplifying, it must be:

$$\nabla_t \times \underline{e}_t = -j\omega\mu\, h_z\, \underline{z}_o.$$

Now, taking the second equation of the (4.2) and vector multiplying by $\underline{z}_o$ on the right side, we obtain:

$$\underline{z}_o \times \frac{\partial \underline{E}_t}{\partial z} \times \underline{z}_o = \frac{\partial \underline{E}_t}{\partial z} = -j\omega\mu\, \underline{H}_t \times \underline{z}_o,$$

then:

$$\frac{dV}{dz}\, \underline{e}_t = -j\omega\mu\, I(z)\, \underline{h}_t \times \underline{z}_o. \tag{4.4}$$

The first of (4.3) becomes:

$$\nabla_t \times \underline{h}_t = 0,$$

and the second becomes:

$$- V(z) \, \underline{z}_o \times \nabla_t h_z + \frac{dI}{dz} \, \underline{z}_o \times \underline{h}_t = j\omega\varepsilon_c \, V(z) \, \underline{e}_t. \tag{4.5}$$

If we assume that $\underline{e}_t = \underline{h}_t \times \underline{z}_o$ (which implies that $\underline{h}_t = \underline{z}_o \times \underline{e}_t$, simply multiplying on the left by $\underline{z}_o$) in (4.4), it follows:

$$\frac{dV}{dz} = -j\omega\mu \, I(z) = -jk_z \frac{\omega\mu}{k_z} I(z),$$

which coincides with the *first transmission line equation*, having defined:

$$Z_S^{TE} = j\omega\mu \qquad Z_c^{TE} = \frac{\omega\mu}{k_z}.$$

On the other hand, starting from (4.5) we get:

$$\frac{dI}{dz} \, \underline{z}_o \times \underline{h}_t = \frac{dI}{dz} \left(-\underline{e}_t \right) = \left(j\omega\varepsilon_c \, \underline{e}_t + \underline{z}_o \times \nabla_t h_z \right) V(z).$$

At this point, similarly to what we have seen before:

$$\frac{dI}{dz} = -jk_z \, Y_c^{TE} \, V(z) = -jk_z \frac{k_z}{\omega\mu} V(z) = -j\frac{k_z^2}{\omega\mu} V(z),$$

and eliminating $V(z)$ from the equation we get:

$$-j\frac{k_z^2}{\omega\mu} \left(-\underline{e}_t \right) = j\omega\varepsilon_c \, \underline{e}_t + \underline{z}_o \times \nabla_t h_z,$$

from which, developing:

$$\underline{e}_t \left(j\omega\varepsilon_c - j\frac{k_z^2}{\omega\mu} \right) = \underline{e}_t \frac{-\omega^2\mu\varepsilon_c + k_z^2}{j\omega\mu} = -\underline{z}_o \times \nabla_t h_z.$$

Now we put $k_t^2 = k^2 - k_z^2 = \omega^2\mu\varepsilon_c - k_z^2$ and then:

$$\underline{e}_t \frac{-k_t^2}{j\omega\mu} = -\underline{z}_o \times \nabla_t h_z,$$

$$\underline{e}_t = \frac{j\omega\mu}{k_t^2} \underline{z}_o \times \nabla_t h_z,$$

i.e. we have just expressed $\underline{e}_t$ as a function of h_z. It follows:

$$\underline{h}_t = \underline{z}_o \times \underline{e}_t = -\frac{j\omega\mu}{k_t^2} \nabla_t h_z,$$

so we have finally expressed $\underline{h}_t$ as a function of h_z, too. The important result we obtained so far is that the whole TE field can be expressed as a function of the only longitudinal component h_z, and in particular as a function of $\nabla_t h_z$.

Now let's see what equation $h_z(q_1, q_2)$ must satisfy. We know, from the Helmholtz equation:

$$\nabla_t^2 H_z + \frac{\partial^2 H_z}{\partial z^2} + k^2 H_z = 0,$$

that becomes, with the assumptions made:

$$V(z) \, \nabla_t^2 h_z + h_z \frac{d^2 V}{dz^2} + k^2 h_z V(z) = 0.$$

Introducing the equations of the lines, i.e.:

$$\frac{d^2 V}{dz^2} = -k_z^2 V(z),$$

the following is obtained:

$$V(z) \, \nabla_t^2 h_z - k_z^2 V(z) \, h_z + k^2 V(z) \, h_z = 0,$$

and eliminating $V(z)$ we get:

$$\nabla_t^2 h_z + k_t^2 h_z = 0,$$

which is a *two-dimensional Helmholtz equation*.

4.1.2 TM Fields

We illustrate now the case of TM fields ($H_z \equiv 0$). The two Maxwell's equations can be dually written:

$$\nabla_t \times \underline{E}_t = 0 \Rightarrow \nabla_t \times \underline{e}_t = 0,$$

$$- \underline{z}_o \times \nabla_t E_z + \underline{z}_o \times \frac{\partial \underline{E}_t}{\partial z} = -j\omega\mu \, \underline{H}_t, \tag{4.6}$$

$$\nabla_t \times \underline{H}_t = j\omega\varepsilon_c \, E_z \, \underline{z}_o, \tag{4.7}$$

$$\underline{z}_o \times \frac{\partial \underline{H}_t}{\partial z} = j\omega\varepsilon_c \, \underline{E}_t. \tag{4.8}$$

The (4.7) can be re-written in the following way, introducing the assumption of separability:

$$I(z) \, \nabla_t \times \underline{h}_t (q_1, q_2) = j\omega\varepsilon_c \, E_z (q_1, q_2, z) \, \underline{z}_o.$$

Given the separated form of the first member, it can be deduced that:

$$E_z (q_1, q_2, z) = e_z (q_1, q_2) \, I(z).$$

So the transverse component of $\underline{H}$ and the longitudinal component of $\underline{E}$ have the same dependence on z. After a simplification, the following can be obtained:

$$\nabla_t \times \underline{h}_t = j\omega\varepsilon_c \, e_z \, \underline{z}_o.$$

Executing a vector product of the (4.8) by $\underline{z}_o$ (on the right side), we have:

$$\underline{z}_o \times \frac{\partial \underline{H}_t}{\partial z} \times \underline{z}_o = \frac{\partial \underline{H}_t}{\partial z} = j\omega\varepsilon_c \, \underline{E}_t \times \underline{z}_o,$$

$$\frac{dI}{dz} \underline{h}_t = j\omega\varepsilon_c \, V(z) \, \underline{e}_t \times \underline{z}_o.$$

Now let us assume again $\underline{h}_t = \underline{z}_o \times \underline{e}_t$ as it was done in the TE case, it follows:

$$\frac{dI}{dz} = -j\omega\varepsilon_c \, V(z) = -jk_z \frac{\omega\varepsilon_c}{k_z} V(z),$$

which after imposing:

$$Y_P^{TM} = j\omega\varepsilon_c \qquad Y_c^{TM} = \frac{\omega\varepsilon_c}{k_z} \Rightarrow Z_c^{TM} = \frac{k_z}{\omega\varepsilon_c}.$$

coincides with the second transmission-line equation.

On the other hand from (4.6) we can obtain:

$$-I(z) \, \underline{z}_o \times \nabla_t e_z + \frac{dV}{dz} \underline{z}_o \times \underline{e}_t = -j\omega\mu \, I(z) \, \underline{h}_t,$$

then:

$$\frac{dV}{dz}\underline{h}_t = \left(-j\omega\mu\,\underline{h}_t + \underline{z}_o \times \nabla_t e_z\right) I(z).$$

Putting now:

$$\frac{dV}{dz} = -jk_z\, Z_c^{TM}\, I(z) = -jk_z\,\frac{k_z}{\omega\varepsilon_c}\, I(z) = -j\frac{k_z^2}{\omega\varepsilon_c}\, I(z),$$

and eliminating $I(z)$, we get the relation:

$$-j\frac{k_z^2}{\omega\varepsilon_c}\,\underline{h}_t = -j\omega\mu\,\underline{h}_t + \underline{z}_o \times \nabla_t e_z,$$

$$\underline{h}_t\,\frac{k_z^2 - \omega^2\mu\varepsilon_c}{j\omega\varepsilon_c} = \underline{h}_t\,\frac{-k_t^2}{j\omega\varepsilon_c} = \underline{z}_o \times \nabla_t e_z,$$

$$\underline{h}_t = -\frac{j\omega\varepsilon_c}{k_t^2}\,\underline{z}_o \times \nabla_t e_z,$$

in which $\underline{h}_t$ is expressed as a function of e_z. Then we have:

$$\underline{e}_t = \underline{h}_t \times \underline{z}_o = -\frac{j\omega\varepsilon_c}{k_t^2}\,\nabla_t e_z$$

and so $\underline{e}_t$ is also expressed as a function of e_z (more precisely, as a function of $\nabla_t e_z$). So the whole TM field can be expressed as a function of the only longitudinal component e_z.

Note that also $e_z(q_1, q_2)$ must satisfy the two-dimensional homogeneous Helmholtz equation. In fact, from:

$$\nabla_t^2 E_z + \frac{\partial^2 E_z}{\partial z^2} + k^2\, E_z = 0,$$

we have, with the assumptions made,

$$I(z)\,\nabla_t^2 e_z + e_z\,\frac{d^2 I}{dz^2} + k^2\, e_z\, I(z) = 0,$$

and so:

$$I(z)\,\nabla_t^2 e_z - k_z^2\, I(z)\, e_z + k^2\, I(z)\, e_z = 0,$$

that can be simplified by removing the common factor $I(z)$ obtaining:

$$\nabla_t^2 e_z + k_t^2\, e_z = 0.$$

Note that an alternative approach which takes advantage of the use of the *vector potential* can be adopted. In particular, a purely longitudinal potential $\underline{A} = A_z \underline{z}_o$ gives rise to a TM$^{(z)}$ field, and therefore A_z results proportional to E_z, while a purely longitudinal potential $\underline{F} = F_z \underline{z}_o$ gives rise to a TE$^{(z)}$ field, and therefore F_z is proportional to H_z.

4.1.3 TEM Fields

Finally, TEM fields ($E_z = H_z \equiv 0$) are considered. The Maxwell's equations become

$$\nabla_t \times \underline{E}_t = 0,$$

$$\nabla_t \times \underline{H}_t = 0,$$

$$\underline{z}_o \times \frac{\partial \underline{E}_t}{\partial z} = -j\omega\mu \, \underline{H}_t,$$

$$\underline{z}_o \times \frac{\partial \underline{H}_t}{\partial z} = j\omega\varepsilon_c \, \underline{E}_t.$$

Using again the conditions on the separability of the transverse fields it is obtained:

$$\nabla_t \times \underline{e}_t = 0,$$

$$\nabla_t \times \underline{h}_t = 0,$$

$$\frac{dV}{dz} \underline{z}_o \times \underline{e}_t = -j\omega\mu \, I(z) \underline{h}_t, \tag{4.9}$$

$$\frac{dI}{dz} \underline{z}_o \times \underline{h}_t = j\omega\varepsilon_c \, V(z) \underline{e}_t. \tag{4.10}$$

The (4.9) becomes the first transmission line equation through the usual position $\underline{e}_t = \underline{h}_t \times \underline{z}_o$:

$$\frac{dV}{dz} = -j\omega\mu \, I(z),$$

while (4.10) becomes:

$$\frac{dI}{dz} = -j\omega\varepsilon_c \, V(z).$$

From the first, recalling the equations of the lines in their two possible forms, it results:

$$Z_S^{\text{TEM}} = j\omega\mu \qquad Z_c^{\text{TEM}} = \frac{\omega\mu}{k_z};$$

from the second it is:

$$Y_P^{\text{TEM}} = j\omega\varepsilon_c \qquad Y_c^{\text{TEM}} = \frac{\omega\varepsilon_c}{k_z}.$$

From the condition $Z_c^{\text{TEM}} \cdot Y_c^{\text{TEM}} = 1$ it follows:

$$\frac{\omega\mu}{k_z}\frac{\omega\varepsilon_c}{k_z} = 1,$$

$$k_z^2 = \omega^2\mu\varepsilon_c \quad \Rightarrow \quad k_z = k \quad \Rightarrow \quad k_t^2 = 0,$$

and it is finally:

$$Z_c^{\text{TEM}} = \sqrt{\frac{\mu}{\varepsilon_c}} = \zeta \qquad Y_c^{\text{TEM}} = \sqrt{\frac{\varepsilon_c}{\mu}} = \frac{1}{\zeta}.$$

This is essentially a characteristic property of TEM waves that has also occurred in the case of uniform plane waves: the propagation constant of a TEM field in a guide coincides with the one of a uniform plane wave that propagates in a free space filled with the same medium.

It should be observed that (4.9) and (4.10) don't give information related to the vectorial part. In order to add this kind of information we need to use different field properties, particularly $\nabla \cdot \underline{D} = \rho$ is used, from which, considering a region of space with no free charges and filled with a homogeneous medium, it follows $\nabla \cdot \underline{E} = 0$. So:

$$\left(\nabla_t + \underline{z}_o \frac{\partial}{\partial z}\right) \cdot \left(\underline{E}_t + E_z \underline{z}_o\right) = 0;$$

and since $E_z = 0$, it follows:

$$\nabla_t \cdot \underline{e}_t = 0.$$

On the other hand we have already seen that:

$$\nabla_t \times \underline{e}_t = 0 \quad \Rightarrow \quad \underline{e}_t = -\nabla_t \Phi(q_1, q_2).$$

$\underline{e}_t$ is both irrotational and solenoidal, therefore we can write:

$$\nabla_t \cdot \underline{e}_t = \nabla_t \cdot (-\nabla_t \Phi) = 0 \quad \Rightarrow \quad \nabla_t^2 \Phi = 0;$$

the latter is a *two-dimensional Laplace equation*.

The same considerations, of course, could be applied to the magnetic field, this time starting from $\nabla \cdot \underline{B} = 0$. Once one field is known, the other follows easily.

4.2 Boundary Conditions for Metallic Waveguides. Propagation Modes

Let us now consider the case of metal waveguides cables, virtually perfect conductors, and let us see how the boundary conditions can be expressed in terms of the formalism introduced above for the three considered cases TE, TM and TEM. Henceforth we will speak about propagation modes to indicate the possible solutions of the differential electromagnetic-field equations which satisfy the boundary conditions imposed by the guiding structure and therefore are effectively able to propagate (even alone) in such a structure.

Let's start from the TE field ($E_z \equiv 0$). The boundary condition requires that $\underline{e}_t \cdot \underline{s}_o = 0$ on the edge s of the cross section S of the guide. Let us recall that the field $\underline{e}_t$ was actually related to:

$$\underline{z}_o \times \nabla_t h_z = 0 \quad \Rightarrow \quad \underline{z}_o \times \nabla_t h_z \cdot \underline{s}_o = \underline{s}_o \times \underline{z}_o \cdot \nabla_t h_z = \underline{n}_o \cdot \nabla_t h_z = 0 \quad su \quad s,$$

then:

$$\frac{\partial h_z}{\partial n} = 0 \quad su \quad s.$$

In the case of TM fields ($H_z \equiv 0$) the condition requires $e_z \equiv 0$ on s, and also $\underline{e}_t \cdot \underline{s}_o = 0$ on s, therefore:

$$\nabla_t e_z \cdot \underline{s}_o = 0 \quad \Rightarrow \quad \frac{\partial e_z}{\partial s} = 0 \quad on \quad s.$$

It is clear, however, that the second condition found is included in the first. The reference system is shown in Fig. 4.1.

Finally, the case of TEM fields ($E_z = H_z \equiv 0$) is considered. The requirement is $\underline{e}_t \cdot \underline{s}_o = 0$ on s and then:

Fig. 4.1 Reference system for cylindrical waveguides

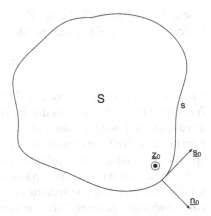

$$(-\nabla_t \Phi) \cdot \underline{s}_o = 0 \quad \Rightarrow \quad \frac{\partial \Phi}{\partial s} = 0 \quad on \quad s,$$

i.e. $\Phi(q_1, q_2)$ is constant on s. On the other hand Φ appeared to be a solution of Laplace equation, and it is known that the solutions of this equation present the maximum and minimum values on the boundary of the relevant domain. But, since Φ results constant on the border, then Φ must be constant over the entire S cross section: it follows that $\nabla_t \Phi \equiv 0$, and as a consequence $\underline{e}_t \equiv 0 \Rightarrow \underline{h}_t \equiv 0$ and so only the identically zero trivial solution is found, of no physical interest. Therefore, if the boundary is a single line, i.e. the section is simply connected, TEM waves cannot propagate; while, if the region is a multi-connected section (for example, a coaxial cable), then the (constant) values of Φ on the various edges can be different from one another and therefore the condition $\Phi = constant$ on the whole S is no longer necessary. So, for example, TEM waves can propagate in a coaxial cable. In particular, in this case only a single TEM wave can propagate: in fact it can be demonstrated that $n - 1$ distinct TEM waves can propagate in a waveguide whose cross-section is n-times connected (i.e., the contour is constituted by n separated parts).

It is worth recalling that the boundary conditions, in general, determine the values of k_t^2 depending on the particular shape and size of the guiding structure. In particular, the differential equation is the same for TE and TM modes, but the solutions can be very different for different boundary conditions; instead, in the TEM case the differential equation itself is different.

4.3 The Guided Propagation as an Eigenvalue Problem

The typical form of a two-dimensional Helmholtz equation for TE and TM modes is the following:

$$\nabla_t^2 T(q_1, q_2) + k_t^2 T(q_1, q_2) = 0,$$

where $T \equiv e_z$ in the TM case, $T \equiv h_z$ in the TE case. This equation assumes the typical form of an eigenvalue problem as $L\varphi = \lambda\varphi$:

$$-\nabla_t^2 T = k_t^2 T,$$

where the L operator is represented by $-\nabla_t^2$, the eigenvalue λ by k_t^2 and the eigenfunction φ by T. It is known that this problem admits non-trivial solutions only for a particular set of k_t^2 values and that at least one eigenfunction T (not identically zero by definition) corresponds to any eigenvalue, and the eigenfunctions are defined apart from a multiplicative arbitrary constant.

It can be shown that the eigenvalues of the $-\nabla_t^2$ operator form a countable infinite set (the so-called discrete spectrum) in the case of closed metal guides, and that the corresponding eigenfunctions represent a complete set, i.e. a representation base for the L^2 vector space of the functions square summable, that is, for the functions whose square modulus is integrable and the integral has finite value. Such representations are in general called spectral representations.

In the case of metal structures with openings, or of dielectric waveguides, the discrete spectrum (guided modes) is joined by a continuous spectrum (radiation modes). The latter eigenfunctions correspond to the possibility that there is emission of energy from the guide towards the outside, as opposite to the guided modes for which the field is evanescent (attenuates) outside the guide.

Here some reminders of functional analysis are helpful. Let us start with some definitions: given any operator L, is called adjoint L^a of L (when it exists) an operator such that:

$$< Lf, g >=< f, L^a g >,$$

where the brackets indicate a scalar product of (in general) complex functions, that is defined in our case as follows:

$$< f, g >= \int_S f\, g^*\, dS,$$

being S the cross section of the waveguide. When $L^a \equiv L$, i.e., when

$$< Lf, g >=< f, Lg >,$$

the operator is said *self-adjoint* or **Hermitian**.

From the definition of scalar product introduced before, and more generally from the axioms of its algebraic structure, it follows that:

$$< f, g >=< g, f >^* .$$

In the case of a self-adjoint operator then, if $f \equiv g$, we have:

$$< Lf, f >=< f, Lf >=< Lf, f >^*,$$

and therefore the quantity $< Lf, f >$ is real, and so it makes sense to study its sign. In particular, if $< Lf, f >$ is always ≥ 0, and $< Lf, f >= 0$ (if and) only if $f \equiv 0$, the operator is said positive definite. Instead, if $< Lf, f >$ may be zero also for functions f not identically zero, the operator is said positive semidefinite. In a similar way is defined the operator negative definite or negative semidefinite.

Note that if the f function is an eigenfunction φ, it is:

$$< L\varphi, \varphi >=< \lambda\varphi, \varphi >= \lambda < \varphi, \varphi > \quad \Rightarrow \quad \lambda = \frac{< L\varphi, \varphi >}{< \varphi, \varphi >}.$$

Therefore, the eigenvalues λ are real if the operator L is self-adjoint. Note in fact that:

$$<\varphi, \varphi> = \int_S \varphi \varphi^* \, dS = \int_S |\varphi|^2 \, dS > 0.$$

Moreover, if the operator L is also positive definite or positive semidefinite, the eigenvalues are > 0 or ≥ 0, respectively.

In addition to having real eigenvalues, a self-adjoint operator has also the property that eigenfunctions corresponding to different eigenvalues are orthogonal to one another, i.e. their scalar product, defined in a vector space of functions, is zero. In fact, given two eigenfunctions φ_i, φ_j corresponding to distinct (real) eigenvalues λ_i, λ_j, it is:

$$< L\varphi_i, \varphi_j > = \lambda_i < \varphi_i, \varphi_j > = < \varphi_i, L\varphi_j > = \lambda_j < \varphi_i, \varphi_j >$$

$$\Downarrow$$

$$(\lambda_i - \lambda_j) < \varphi_i, \varphi_j > = 0,$$

so:

$$\lambda_i \neq \lambda_j \quad \Rightarrow \quad < \varphi_i, \varphi_j > = 0.$$

In the case of multiple eigenvalues, i.e. eigenvalues corresponding to $n > 1$ distinct (linearly independent) eigenfunctions, such eigenfunctions (which are called degenerate) are not necessarily orthogonal. However, it is possible to apply an algebraic procedure called *Gram-Schmidt orthogonalization*, which permits to pass from these n eigenfunctions to a new set of n eigenfunctions that, instead, are mutually orthogonal (and then, through the division by their modulus $\sqrt{< \varphi, \varphi >}$ they can become orthonormal, i.e. orthogonal with unitary modulus). The procedure uses linear combinations, and of course such combinations are also eigenfunctions associated to the same eigenvalue. In conclusion, then, if the operator is self-adjoint it is possible to find a basis for the function space consisting of (infinite, in general) orthogonal eigenfunctions.

The advantage of having an orthogonal basis (with respect to a base simply constituted by linearly independent functions) resides in the immediate possibility to compute the coefficients of the linear combination. It is in fact, for a generic function of the space (recalling that we are dealing with infinite-dimensional spaces):

$$f(P) = \sum_{n=1}^{\infty} c_n \varphi_n(P).$$

Multiplying both sides by the scalar function $\varphi_m(P)$, it follows:

$$< f, \varphi_m > = < \sum_{n=1}^{\infty} c_n \varphi_n, \varphi_m > = \sum_{n=1}^{\infty} c_n < \varphi_n, \varphi_m > =$$

$$= \sum_{n=1}^{\infty} c_n \delta_{nm} = c_m \quad m = 1, 2, \ldots \infty,$$

having used the orthonormality in the last step and having introduced the Kronecker symbol:

$$< \varphi_n, \varphi_m > = \delta_{nm} = \begin{cases} 1 & \text{if } m = n \\ 0 & \text{if } m \neq n \end{cases}.$$

The coefficients c_n are generally called Fourier coefficients as the expression seen generalizes the formula which describes the Fourier series.

Using the premises shown above, we can now demonstrate that $-\nabla_t^2$ operator is self-adjoint for boundary conditions relevant to perfectly conducting walls. Let us remember first the Green's lemma in its second form:

$$\oint_S \left(f \frac{\partial g}{\partial n} - g \frac{\partial f}{\partial n} \right) dS = \int_V \left(f \nabla^2 g - g \nabla^2 f \right) dV,$$

where S is, as usual, the closed surface that bounds the volume V, and where the direction of the normal $\underline{n}$ is outgoing the volume. This formula must be particularized to our two-dimensional space, obtaining:

$$\oint_s \left(f \frac{\partial g}{\partial n} - g \frac{\partial f}{\partial n} \right) ds = \int_S \left(f \nabla_t^2 g - g \nabla_t^2 f \right) dS,$$

being s a closed line which bounds the open surface S, which represents the cross section of the guide, while ∇ is replaced by ∇_t.

The operator L is self-adjoint if the following requirement is met:

$$< Lf, g > - < f, Lg > = 0,$$

i.e. in our case:

$$< -\nabla_t^2 f, g > - < f, -\nabla_t^2 g > = \int_S \left(-\nabla_t^2 f g^* + f \nabla_t^2 g^* \right) dS =$$

$$= \oint_s \left(f \frac{\partial g^*}{\partial n} - g^* \frac{\partial f}{\partial n} \right) ds,$$

having applied the Green's lemma to the functions f and g^*, and ∇_t^2 being a real operator (i.e. when operating on a real function, it yields another real function).

Let us observe now that, in our application to the perfectly conducting waveguides, the functions f and g^*, to which the operator $-\nabla_t^2$ is applied, must satisfy the boundary conditions for TE and TM modes, for example, with reference to the function f:

$$\frac{\partial f}{\partial n} = 0 \quad on \quad s \quad , \quad f = 0 \quad on \quad s,$$

respectively. So, in these two cases, the circulation is zero and the operator results self-adjoint.

This example shows us clearly that is not an intrinsic property for the operator to be self-adjoint (or not), but it depends on the particular boundary conditions, which actually determine the so-called *domain* of the operator itself, i.e. the particular set of functions on which it operates.

From the fact that the eigenvalues $\lambda = k_t^2$ are real, it follows that the eigenfunctions T can be considered real. In fact, a generic complex eigenfunction can always be written $T = T_R + jT_J$, where T_R and T_J are real. But then, by separating the real and imaginary parts of the differential equation and knowing that k_t^2 is real, it follows:

$$\nabla_t^2 T_R + k_t^2 T_R = 0 \quad , \quad \nabla_t^2 T_J + k_t^2 T_J = 0.$$

Moreover, boundary conditions also hold separately for the real and the imaginary part for the various types of modes. So T_R and T_J are separately eigenfunctions related to the same eigenvalue, and the eigenfunction T can be regarded as their linear combination.

On the other hand if the eigenfunction is real, and therefore has a constant phase, the variability of the phase resides only in the $V(z)$ and $I(z)$ longitudinal functions, and so the $z = constant$ planes are equiphase surfaces, on which, however, the field amplitude generally changes. It follows that the modes of the waveguides with perfectly conducting walls are non-uniform plane waves.

It should be noted, finally, that, in the case of the closed metal guides filled with a homogeneous material, the k_t^2 eigenvalues and the T eigenfunctions depend only on the geometry (shape and size) of the guide cross section and they are independent of frequency and of the medium parameters (ε, μ, σ).

We can now observe that the operator $-\nabla_t^2$ is substantially positive definite when assuming those boundary conditions. We need to analyze the sign, as already seen, of the quantity:

$$< Lf, f > = < -\nabla_t^2 f, f > = \int_S -f^* \nabla_t^2 f \, dS.$$

This time we apply the first form of Green's lemma which was already presented in its three-dimensional version reported here again for simplicity:

$$\oint_S f \frac{\partial g}{\partial n} \, dS = \int_V \left(\nabla f \cdot \nabla g + f \nabla^2 g \right) dV.$$

In the two-dimensional case the Green's lemma becomes:

$$\oint_s f \frac{\partial g}{\partial n} ds = \int_S \left(\nabla_t f \cdot \nabla_t g + f \nabla_t^2 g \right) dS.$$

Now we can apply the lemma to the f^* and f functions, obtaining:

$$\int_S -f^* \nabla_t^2 f \, dS = \int_S \nabla_t f^* \cdot \nabla_t f \, dS - \oint_s f^* \frac{\partial f}{\partial n} ds = \int_S |\nabla_t f|^2 \, dS,$$

since the boundary integral is null for perfectly conducting wall boundary conditions.
Note now that the quantity:

$$\int_S |\nabla_t f|^2 \, dS = < Lf, f > \text{ results } \geq 0$$

We must now find the condition for which it is zero. This happens if and only if $\nabla_t f \equiv 0$ on S, i.e. $f = constant$ on S. Hence, in the case of TM modes it must be $f \equiv 0$ on S for the continuity of f, since we saw that in this case $f = 0$ on s; therefore the operator is positive definite. Instead in the case of TE modes, from:

$$\frac{\partial f}{\partial n} = 0 \quad on \quad s,$$

it does not necessarily follow that $f \equiv 0$ on S, and therefore the operator is positive semidefinite. The solution $T = constant$ on S is an eigenfunction associated to the eigenvalue $k_t^2 = 0$, as can be seen from

$$\lambda = \frac{< L\varphi, \varphi >}{< \varphi, \varphi >}.$$

This solution, however, is only possible when the electromagnetic field is identically zero, because it is $\nabla_t h_z \equiv 0$ in the expressions for the components of the transverse field. Therefore the domain of the operator could be redefined in order to exclude the constant solutions, so that the operator becomes positive definite on this new domain.

The direct consequence is that the eigenvalues $\lambda = k_t^2$ are real and positive, in confirmation of their formal quadratic expression.

After reviewing the transverse eigenvalue problem, we are going to analyze the propagation features in the longitudinal z direction. Let us assume that the medium filling the guide is non-dispersive and non-dissipative. In this case, the separability condition implies:

$$k_z^2 = k^2 - k_t^2 = \omega^2 \mu \varepsilon - k_t^2,$$

which is a real quantity, therefore

$$k_z = \sqrt{\omega^2 \mu \varepsilon - k_t^2}$$

can be either purely real if $\omega^2 \mu \varepsilon > k_t^2$, or purely imaginary if $\omega^2 \mu \varepsilon < k_t^2$. The first case corresponds to a mode propagation with no attenuation along z and with a behavior of the type $e^{-j\beta_z z}$. The second case corresponds to an exponential attenuation of the field along z with no propagation, i.e., with no phase change, and so its behavior is expressed by $e^{-\alpha_z z}$. The two cases are separated by a particular value of ω, called ω_c (cut-off angular frequency), such that:

$$\omega_c^2 \mu \varepsilon = k_t^2 \quad \Rightarrow \quad \omega_c = \frac{k_t}{\sqrt{\mu \varepsilon}} = k_t \, v.$$

For $\omega = \omega_c$ it is $k_z = 0$. The *cut-off frequency*:

$$f_c = \frac{\omega_c}{2\pi},$$

and *cut-off wavelength*:

$$\lambda_c = \frac{2\pi}{k_t},$$

are also defined to describe the cut-off condition. Note that each mode has its own eigenvalue k_t^2, and therefore its own cut-off frequency, but it may happen that two (degenerate) eigenfunctions have the same eigenvalue and therefore the two corresponding (degenerate) modes have the same cut-off frequency.

Propagation without attenuation occurs when $\omega > \omega_c$, while for $\omega < \omega_c$ there is attenuation with no propagation, it follows that the waveguide behaves as a high-pass filter. The mode with the lowest eigenvalue, and therefore with the lowest cut-off frequency, is called the *dominant* or *fundamental* mode; all other modes are called *higher-order modes* or simply *higher modes*. To understand the reason for such name let us suppose to operate the waveguide in a frequency band between the cut-off frequency of the dominant mode and the cut-off frequency of the second mode, i.e. the first higher-order mode. It is clear that the dominant mode will propagate with no attenuation in this band, while all higher modes will be attenuated and they won't propagate. Assuming that a generator excites in the waveguide a generic field (that can always be expressed as a series expansion of the complete set of modes), it will happen that at a certain distance from the section of the excitation, the electromagnetic field will be in practice represented by the only term associated with the dominant mode. In the case in which the cylindrical structure of the waveguide is perturbed by the presence of an obstacle (discontinuity, bend), all the infinite higher attenuated modes will again appear in the vicinity of such obstacle, but at a certain distance they will be again negligible.

So a single transmission line, corresponding to the dominant mode, can be associated to the waveguide of our example, while any effect due to the discontinuity is "concentrated" near the discontinuity itself and therefore representable, in the equivalent circuit, as a concentrate component, i.e. an impedance (or admittance) that will be purely imaginary, i.e. of reactive type, in the absence of losses and that will take into account the reactive power of all attenuated modes, concentrated in the vicinity of the obstacle.

Note also that, if the guiding structure supports a TEM mode, this is certainly the dominant mode because it is characterised by $k_t^2 = 0$ and therefore by $f_c = 0$. The field can then propagate in such structures up to arbitrarily low frequencies, so it propagates even in fairly static conditions.

A guide wavelength:

$$\lambda_z = \frac{2\pi}{\beta_z},$$

and a *guide phase velocity*:

$$v_z = \frac{\omega}{\beta_z},$$

are defined in a waveguide; these quantities are related to those of free space (filled with the same medium that fills the guide) by the following relations:

$$\lambda_z = \frac{2\pi}{\beta_z} = \frac{2\pi}{\sqrt{k^2 - k_t^2}} = \frac{2\pi}{k\sqrt{1 - \frac{k_t^2}{\omega^2 \mu \varepsilon}}} = \frac{\lambda}{\sqrt{1 - \left(\frac{f_c}{f}\right)^2}},$$

$$v_z = \frac{\omega}{\beta_z} = \frac{\omega}{\omega\sqrt{\mu \varepsilon}\sqrt{1 - \left(\frac{f_c}{f}\right)^2}} = \frac{v}{\sqrt{1 - \left(\frac{f_c}{f}\right)^2}}.$$

Note that for $f > f_c$, i.e. in propagation, it is $\lambda_z > \lambda$ and $v_z > v$ (fast wave). Finally, the group velocity in the waveguide is expressed by:

$$v_{gz} = \frac{1}{\frac{d\beta_z}{d\omega}}.$$

On the other hand, we have:

$$\frac{d\beta_z}{d\omega} = \frac{d}{d\omega}\sqrt{\omega^2 \mu \varepsilon - k_t^2} = \frac{\omega \mu \varepsilon}{\sqrt{\omega^2 \mu \varepsilon - k_t^2}} = \frac{\mu \varepsilon}{\sqrt{\mu \varepsilon}\sqrt{1 - \left(\frac{f_c}{f}\right)^2}} = \frac{1}{v\sqrt{1 - \left(\frac{f_c}{f}\right)^2}},$$

then:

$$v_{gz} = v \sqrt{1 - \left(\frac{f_c}{f}\right)^2},$$

it is $v_{gz} < v$, as it should be, in case of propagation.

Now let us assume that the medium filling the guide is still non-dispersive (ε and μ real), but dissipative ($\sigma \neq 0$), the propagation constant k_z will certainly be complex and it will be:

$$k_z^2 = \omega^2 \mu \varepsilon_c - k_t^2 = \omega^2 \mu \varepsilon - j\omega\mu\sigma - k_t^2 = \left(\omega^2 \mu \varepsilon - k_t^2\right) - j\omega\mu\sigma.$$

It can be seen that the complex number k_z^2 belongs to the fourth quadrant of the complex plane for $f > f_c$, while it belongs to the third for $f < f_c$. If we extract the square root with the usual determination $k_{z\,R} > 0$, it results in both cases that k_z belongs to the fourth quadrant. So one can always put $k_z = \beta_z - j\alpha_z$ being β_z and α_z both positive. Moreover, it results that for $f > f_c$, $\beta_z > \alpha_z$, while for $f < f_c$ the attenuation prevails as is reasonable, i.e. $\alpha_z > \beta_z$. At the cut-off condition (which, however, no longer corresponds to a sharp cut) it is $k_z^2 = -j\omega\mu\sigma$, so k_z lies on the bisector of the second and fourth quadrants ($\beta_z = \alpha_z$).

4.4 Rectangular Waveguide

We are going to consider now a rectangular waveguide of internal dimensions a and b as a first simple example. In this case, it is convenient to work in xy Cartesian coordinates to facilitate the imposition of the boundary conditions; moreover, the differential equation is easier to solve in rectangular coordinates. It is therefore:

$$T(q_1, q_2) = T(x, y),$$

and the two-dimensional Helmholtz equation becomes:

$$\frac{\partial^2 T}{\partial x^2} + \frac{\partial^2 T}{\partial y^2} + k_t^2 T = 0.$$

We are going to use the separation of variables as a solution method, as we already did for plane waves. So, assuming:

$$T(x, y) = X(x) Y(y),$$

and substituting it in the equation:

$$Y(y) \frac{d^2 X}{dx^2} + X(x) \frac{d^2 Y}{dy^2} + k_t^2 X(x) Y(y) = 0,$$

and finally dividing the three addends by the term $X(x) Y(y)$ which is clearly not identically zero we obtain:

$$\frac{1}{X(x)} \frac{d^2 X}{dx^2} + \frac{1}{Y(y)} \frac{d^2 Y}{dy^2} + k_t^2 = 0.$$

In the above equation the first term depends only on x, the second depends only on y, and the third is independent of both x and y. Then, by deriving the equation with respect to x and y, we obtain:

$$\frac{d}{dx} \left[\frac{1}{X(x)} \frac{d^2 X}{dx^2} \right] = 0,$$

$$\frac{d}{dy} \left[\frac{1}{Y(y)} \frac{d^2 Y}{dy^2} \right] = 0,$$

from which:

$$\frac{1}{X(x)} \frac{d^2 X}{dx^2} = constant = -k_x^2 \Rightarrow \frac{d^2 X}{dx^2} + k_x^2 X(x) = 0,$$

$$\frac{1}{Y(y)} \frac{d^2 Y}{dy^2} = constant = -k_y^2 \Rightarrow \frac{d^2 Y}{dy^2} + k_y^2 Y(y) = 0,$$

where the constants are in general complex, and cannot be completely arbitrary as usual, having to satisfy the condition of separability:

$$k_x^2 + k_y^2 = k_t^2.$$

In particular, these constants cannot be both zero, otherwise we would have $k_t^2 = 0$ and so the TEM mode, which cannot exist since the section is simply connected.

Assuming $k_x \neq 0$ in the first equation, the general solution (written in the form of standing waves, which is convenient in limited structures):

$$X(x) = C_1 \sin(k_x x) + C_2 \cos(k_x x).$$

Similarly, assuming $k_y \neq 0$ in the second equation, it is:

$$Y(y) = D_1 \sin(k_y y) + D_2 \cos(k_y y).$$

Instead $X(x) = C_1 x + C_2$ and, similarly, $Y(y) = D_1 y + D_2$ follow respectively from $k_x = 0$ and $k_y = 0$.

Let us now impose the boundary conditions in the presence of perfectly conducting walls, starting from the TM case, for which the condition $T = 0$ on s holds. The four conditions $X(0) = 0$, $X(a) = 0$, $Y(0) = 0$, $Y(b) = 0$ must be imposed on $x = 0$, $x = a$, $y = 0$, $y = b$ respectively. From the first condition, and in the case $k_x \neq 0$, it follows:

$$C_2 = 0 \Rightarrow X(x) = C_1 \sin(k_x x).$$

Imposing the second condition we get:

$$C_1 \sin(k_x a) = 0 \Rightarrow k_x a = m\pi,$$

$$\Rightarrow k_x = \frac{m\pi}{a} \quad m = 1, 2, 3, \ldots \quad .$$

In the case $k_x = 0$ it is, instead, $C_2 = 0$ from the first condition, and then $C_1 a = 0 \Rightarrow C_1 = 0$ for the second condition; this leads to the trivial solution identically zero. The third and fourth conditions are treated in a similar manner.

So the following expressions were found for TM modes:

$$X(x) = C_1 \sin\left(\frac{m\pi}{a}x\right) \quad m = 1, 2, 3, \ldots,$$

$$Y(y) = D_1 \sin\left(\frac{n\pi}{b}y\right) \quad n = 1, 2, 3, \ldots,$$

and in conclusion the following eigenfunction is found for TM modes:

$$T(x, y) = C \sin\left(\frac{m\pi}{a}x\right) \sin\left(\frac{n\pi}{b}y\right) \quad m, n = 1, 2, 3, \ldots,$$

associated to the eigenvalue:

$$k_t^2 = \left(\frac{m\pi}{a}\right)^2 + \left(\frac{n\pi}{b}\right)^2.$$

Note that the eigenvalues are real and positive, as it is in general demonstrated for metal guides. The remaining arbitrary constant C, typical of the homogeneous problems, can be determined using a suitable normalization condition.

The generic modal solution is then characterised by the pair (m, n) of indexes, for this reason we usually talk about TM_{mn} mode. Let us now impose the boundary conditions for the TE modes. On the $x = 0$ side we must have:

$$-\frac{dX}{dx}\bigg|_{x=0} = 0;$$

on the $x = a$ side it must be:

$$\frac{dX}{dx}\bigg|_{x=a} = 0;$$

on the $y = 0$ side it must be:

$$-\frac{dY}{dy}\bigg|_{y=0} = 0;$$

and in conclusion on the $y = b$ side it has to be:

$$\frac{dY}{dy}\bigg|_{y=b} = 0.$$

Assuming that $k_x \neq 0$, then:

$$\frac{dX}{dx} = C_1 k_x \cos(k_x x) - C_2 k_x \sin(k_x x).$$

From the first condition it follows:

$$C_1 k_x = 0 \Rightarrow C_1 = 0,$$

being $k_x \neq 0$ for hypothesis. At this point, from the second condition it follows:

$$-C_2 k_x \sin(k_x a) = 0 \Rightarrow k_x = \frac{m\pi}{a} \quad m = 1, 2, 3, \ldots \Rightarrow X(x) = C_2 \cos\left(\frac{m\pi}{a}x\right).$$

Let us consider now the case in which $k_x = 0$:

$$\frac{dX}{dx} = C_1 \Rightarrow C_1 = 0 \Rightarrow X(x) = C_2.$$

The above solution may be unified in the:

$$X(x) = C_2 \cos\left(\frac{m\pi}{a}x\right),$$

permitting also the $m = 0$ value.

In a similar way, imposing the boundary conditions on $Y(y)$, we get:

$$Y(y) = D_2 \cos\left(\frac{n\pi}{b}y\right) \quad k_y = \frac{n\pi}{b} \quad n = 0, 1, 2, 3, \ldots.$$

It should be noted, however, that k_x and k_y cannot be both zero having to be, as already seen, $k_x^2 + k_y^2 = k_t^2 \neq 0$; so also m and n cannot be both zero.

The eigenfunction results:

$$T(x, y) = C \cos\left(\frac{m\pi}{a}x\right) \cos\left(\frac{n\pi}{b}y\right),$$

associated to the eigenvalue

$$k^2_{t\,[mn]} = \left(\frac{m\pi}{a}\right)^2 + \left(\frac{n\pi}{b}\right)^2,$$

corresponding to the longitudinal wave number:

$$k_{z\,[mn]} = \sqrt{k^2 - \left(\frac{m\pi}{a}\right)^2 + \left(\frac{n\pi}{b}\right)^2}.$$

Here the mode is characterized by two indices and is denoted as TE$_{mn}$. Note that eigenvalues have the same expression for both TM and TE modes, so the TM$_{mn}$ and the TE$_{mn}$ have the same eigenvalue, i.e. they are degenerate.

It can be shown that in general the eigenvalues of a guide, having generic cross-section as long as limited, form a countable set superiorly unlimited for both mode types. Considering now the modes TE$_{m0}$ we have:

$$k^2_{t\,[m0]} = \left(\frac{m\pi}{a}\right)^2 \Rightarrow k_{t\,[m0]} = \frac{m\pi}{a}.$$

The interval between this value and the next one (which is $\frac{(m+1)\pi}{a}$) is equal to π/a, so it decreases as a increases. Basically for $a \to \infty$ the structure tends to the so-called metallic parallel-plate guide and the eigenvalues are transformed from a countable infinity to a continue infinity. It is possible to demonstrate also that for a guide with any cross-section limited and simply connected, the dominant mode is always a TE mode.

The smallest TM eigenvalue is for TM$_{11}$ mode and its value is:

$$k^2_{t\,[11]} = \left(\frac{\pi}{a}\right)^2 + \left(\frac{\pi}{b}\right)^2.$$

The smallest eigenvalue for TE modes corresponds instead to TE$_{01}$ or TE$_{10}$. In particular, the dominant mode is TE$_{10}$ if $a > b$ (as it is generally assumed): the eigenvalue is

$$k^2_{t\,[10]} = \left(\frac{\pi}{a}\right)^2,$$

and the longitudinal wave number is:

$$k_{z\,[10]} = \sqrt{k^2 - \left(\frac{\pi}{a}\right)^2}.$$

In the case of a lossless dielectric the cut-off angular frequency is:

$$\omega_{c\,[10]} = \frac{\pi}{a\,\sqrt{\mu\varepsilon}};$$

so the cut-off frequency is:

$$f_{c[10]} = \frac{1}{2a\sqrt{\mu\varepsilon}} = \frac{v}{2a};$$

and the cut-off wavelength is $\lambda_{c[10]} = 2a$. The propagation condition $f > f_{c[10]}$ can be also written $\lambda < \lambda_{c[10]}$ i.e. $\lambda < 2a \Rightarrow a > \frac{\lambda}{2}$. In essence, a guide doesn't contain any mode if it is too narrow and it contains more modes as it becomes larger (increasing the size has the same effect as increasing the frequency, from this point of view).

The eigenfunction of the dominant mode is:

$$T(x, y) = T(x) = C \, \cos\left(\frac{\pi}{a}x\right) = h_z(x),$$

from which it follows:

$$\nabla_t h_z = \underline{x}_o \frac{dh_z}{dx} = -\underline{x}_o \, C \frac{\pi}{a} \, \sin\left(\frac{\pi}{a}x\right) \quad \propto \quad \underline{h}_t = h_x \, \underline{x}_o,$$

$$\nabla_t h_z \times \underline{z}_o = \underline{y}_o \, C \frac{\pi}{a} \, \sin\left(\frac{\pi}{a}x\right) \quad \propto \quad \underline{e}_t = e_y \, \underline{y}_o.$$

The configuration of the fields in the guide section has been therefore determined (Fig. 4.2).

It is important to highlight that such shape does not change varying frequency, and therefore it does not change even under the cut-off condition.

Finally, we can easily show that along z the propagation of the TE$_{10}$ mode is equivalent to the one of two uniform plane waves traveling obliquely in the guide and having wave vector $\underline{k}$ lying on the zx horizontal plane (being $k_y = 0$) and forming an angle θ with the x axis, having then $k_x = k_t = \frac{\pi}{a} = k \cos\theta$, $k_z = k \sin\theta$. These waves are totally reflected from the vertical metal walls present at $x = 0$ and $x = a$, transforming each into the other (Fig. 4.3).

Fig. 4.2 Configuration of the transverse electric field in rectangular waveguide for the mode TE$_{10}$

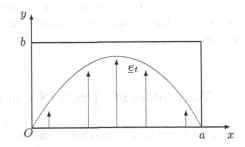

Fig. 4.3 Plane-wave
decomposition of mode TE$_{10}$
in rectangular waveguide

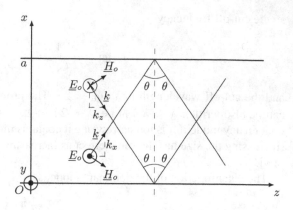

The angle of incidence θ is given by:

$$\tan \theta = \frac{k_z}{k_x} = \frac{\sqrt{k^2 - \left(\frac{\pi}{a}\right)^2}}{\frac{\pi}{a}} = \sqrt{\left(\frac{f}{f_c}\right)^2 - 1} =$$

$$= \sqrt{\left(\frac{f\,2a}{v}\right)^2 - 1} = \sqrt{\left(\frac{2a}{\lambda}\right)^2 - 1},$$

so it depends on the frequency and on the width a of the guide. Hence θ increases with the increase of the frequency or the guide width (the tangent is a monotonically increasing function), and when $f \to \infty$ it follows $\theta \to \pi/2$ i.e. the wave incidence tends to the so-called grazing incidence in the axial direction, analogous to the wave propagation in free space ($k_x = \frac{\pi}{a} \ll k_z, k_z \to k$), in other words it tends to a pure traveling wave along z. Instead, when $f = f_c$, i.e. when we are at cut-off, it is $\theta = 0$, $k_x = k$ and $k_z = 0$, thus there is no longer propagation along z, but a pure standing wave along x. A progressive component along z and a stationary component along x are present instead in all intermediate cases.

These considerations could be extended to all TM and TE modes propagating in a rectangular waveguide, and they demonstrate that the propagation angles of the equivalent plane waves inside the guides cannot be arbitrary, but they need to constitute a discrete set, so that each mode has its own angle, corresponding to its eigenvalue.

4.5 Waveguides of Circular Section

Let us now consider two metallic structures, the circular waveguide and the coaxial cable, characterized by (cylindrical) circular geometry. In this case it is convenient to work in polar coordinates $q_1 = \rho$, $q_2 = \varphi$, to have an easier boundary conditions imposition on the circumferences. So, it is: $T(q_1, q_2) = T(\rho, \varphi)$. Note that the considerations that follow, until explicit imposition is made of the boundary conditions, are also valid in dielectric structures, such as optical fibers, consisting of a central kernel (*core*) and a surrounding mantle (*cladding*).

We need to recall first the expression of the transverse Laplacian in polar coordinates, which is:

$$\nabla_t^2 = \frac{1}{\rho} \frac{\partial}{\partial \rho} \left(\rho \frac{\partial}{\partial \rho} \right) + \frac{1}{\rho^2} \frac{\partial^2}{\partial \varphi^2}.$$

The Helmholtz differential equation is thus:

$$\frac{1}{\rho} \frac{\partial}{\partial \rho} \left(\rho \frac{\partial T}{\partial \rho} \right) + \frac{1}{\rho^2} \frac{\partial^2 T}{\partial \varphi^2} + k_t^2 T = 0.$$

We apply again the method of separation of variables, seeking solutions of the type $T(\rho, \varphi) = P(\rho) \phi(\varphi)$. By inserting this solution in the previous equation, it becomes:

$$\frac{\phi}{\rho} \frac{d}{d\rho} \left(\rho \frac{dP}{d\rho} \right) + \frac{P}{\rho^2} \frac{d^2 \phi}{d\varphi^2} + k_t^2 P \phi = 0.$$

Multiplying both sides by $\frac{\rho^2}{P\phi}$, where neither the numerator nor the denominator can be identically zero, we get:

$$\frac{\rho}{P} \frac{d}{d\rho} \left(\rho \frac{dP}{d\rho} \right) + \frac{1}{\phi} \frac{d^2 \phi}{d\varphi^2} + k_t^2 \rho^2 = 0,$$

where the first and third addend depend only on the ρ variable while the second depends only on the φ variable. Deriving the obtained expression with respect to φ it is obtained:

$$\frac{d}{d\varphi} \left(\frac{1}{\phi} \frac{d^2 \phi}{d\varphi^2} \right) = 0 \Rightarrow \frac{1}{\phi} \frac{d^2 \phi}{d\varphi^2} = constant = -k_\varphi^2 \Rightarrow \frac{d^2 \phi}{d\varphi^2} + k_\varphi^2 \phi = 0.$$

As usual, if $k_\varphi \neq 0$, the general solution $\phi(\varphi) = C_1 \sin(k_\varphi\varphi) + C_2 \cos(k_\varphi\varphi)$ is obtained. Instead when $k_\varphi = 0$ we have $\phi(\varphi) = C_1 \varphi + C_2$. At this point we observe that the $\phi(\varphi)$ function must be periodic of period 2π for the geometric meaning of the variable φ (increasing it of 2π we come back to the same point in the plane). This implies that from $k_\varphi = 0$ it follows necessarily $C_1 = 0$ and $\phi(\varphi) = C_2$. When $k_\varphi \neq 0$ instead, the period of the sinusoidal functions, i.e. $\frac{2\pi}{k_\varphi}$, has to be 2π or a submultiple of 2π, so it must be $k_\varphi = n = 1, 2, 3, \Rightarrow \phi(\varphi) = C_1 \sin(n \varphi) + C_2 \cos(n \varphi)$. If n is allowed to assume the zero value, too, it must be $\phi(\varphi) = C_2$ and then we can include the solution for $k_\varphi = 0$ (these are modes independent of φ, that show interesting properties in terms of loss in the guide, i.e. when the guide walls are not perfectly conductive).

At this point we step back to the differential equation and we insert the constant $-n^2$ in place of the second addend, obtaining:

$$\frac{\rho}{P}\frac{d}{d\rho}\left(\rho\frac{dP}{d\rho}\right) - n^2 + k_t^2\,\rho^2 = 0.$$

Now let us multiply by $P(\rho)$ and perform the derivative:

$$\rho\frac{dP}{d\rho} + \rho^2\frac{d^2P}{d\rho^2} + \left(k_t^2\,\rho^2 - n^2\right)P = 0.$$

Assuming $k_t^2 \neq 0$ (so the TEM mode which may exist in the coaxial cable, but not in the circular waveguide isn't considered, and it will be considered separately) and making the assumption $\xi = k_t\,\rho \Rightarrow \rho = \xi/k_t$:

$$\frac{dP}{d\rho} = \frac{dP}{d\xi}\frac{d\xi}{d\rho} = k_t\frac{dP}{d\xi},$$

$$\frac{d^2P}{d\rho^2} = \frac{d}{d\rho}\left(\frac{dP}{d\rho}\right) = \frac{d}{d\rho}\left(k_t\frac{dP}{d\xi}\right) = k_t\frac{d}{d\xi}\left(k_t\frac{dP}{d\xi}\right) = k_t^2\frac{d^2P}{d\xi^2}.$$

Executing all the proper substitutions in the equation we get:

$$\xi\frac{dP}{d\xi} + \xi^2\frac{d^2P}{d\xi^2} + (\xi^2 - n^2)\,P = 0,$$

which dividing by ξ^2 becomes

$$\frac{d^2P}{d\xi^2} + \frac{1}{\xi}\frac{dP}{d\xi} + \left(1 - \frac{n^2}{\xi^2}\right)P = 0.$$

This well-known differential equation, having variable coefficients, is called Bessel's equation of order n ($n = 0, 1, 2, \ldots$). Like any second-order differential equation, its general solution can be expressed as a linear combination of two independent particular solutions: one possible choice is represented by the so-called Bessel functions of order n of the first kind $J_n(\xi)$ and of the second kind $Y_n(\xi)$. It is therefore

$$P(\rho) = D_1\,J_n(k_t\rho) + D_2\,Y_n(k_t\rho).$$

The following figures represent the Bessel functions behavior (Figs. 4.4 and 4.5). They asymptotically, i.e. for large values of the argument, are similar to damped sinusoidal functions and in fact they are the standing-wave solutions of the Helmholtz equation in cylindrical coordinates:

Fig. 4.4 Bessel J_n functions

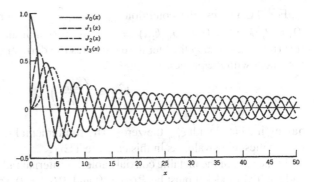

Fig. 4.5 Bessel Y_n functions

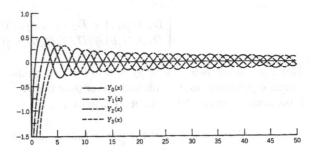

We need to observe that the circular guide section includes the origin $\rho = 0$; all the $Y_n(k_t\rho)$ diverge at the origin, and since the values of the function T are assumed to be limited at any point of the section for physical reasons, we need to impose $D_2 = 0 \Rightarrow P(\rho) = D_1 J_n(k_t\rho)$ for the circular waveguide. In the case of metallic coaxial cable, however, the origin is not part of the domain of interest, then both functions appear. The same observations apply respectively to the core and cladding of the optical fiber.

4.5.1 TE and TM Modes in Circular Metallic Waveguides

Let's impose the boundary conditions for the circular metallic guide having radius a. For TM modes it must be $T = 0$ on s:

$$\phi(\varphi)P(a) = 0 \Rightarrow P(a) = 0 \Rightarrow J_n(k_ta) = 0.$$

The roots of the Bessel function $J_n(\xi)$ must be known to solve the equation. There is an infinite countable number of roots for $J_n(\xi)$, denoted with ξ_{nm} where $n = 0, 1, 2, \ldots; m = 1, 2, 3, \ldots \Rightarrow k_{t_{nm}}^2 = \left(\frac{\xi_{nm}}{a}\right)^2.$

For TE modes, the condition $\frac{\partial T}{\partial n} = \frac{\partial T}{\partial \rho} = 0$ for $\rho = a \Rightarrow \frac{dP}{d\rho}\big|_{\rho=a} = D_1 k_t J_n'(k_t a) = 0 \Rightarrow J_n'(k_t a) = 0$ leads to search all zeros of the $J_n'(\xi)$ derivative (note that usually the literature indicates the derivative with respect to the whole argument with the prime).

$$k_{t_{nm}}^2 = \left(\frac{\xi_{nm}'}{a}\right)^2,$$

having indicated with ξ_{nm}' the zeros. The fundamental mode, which corresponds to the smallest eigenvalue, is in this case the TE$_{11}$.

In the case of the metallic coaxial cable (of internal radius a and external radius b) and for TM modes it must be $P(a) = 0$ and $P(b) = 0$, so:

$$\begin{cases} D_1 J_n(k_t a) + D_2 Y_n(k_t a) = 0 \\ D_1 J_n(k_t b) + D_2 Y_n(k_t b) = 0 \end{cases}.$$

This is a homogeneous linear system in the two unknowns D_1, D_2, and as such it admits eigensolutions, i.e. solutions other than the trivial $D_1 = D_2 = 0$, if and only if the determinant of the coefficients is zero, i.e. if

$$J_n(k_t a) \, Y_n(k_t b) - Y_n(k_t a) \, J_n(k_t b) = 0.$$

This procedure has a very general applicability and the equation obtained is called the characteristic equation of the structure, or dispersion equation (because it allows to obtain the eigenvalues k_t^2, and then $k_z = \sqrt{k^2 - k_t^2}$ as a function of frequency). The cases previously presented are very simple, actually to obtain the eigenvalues we need usually to numerically solve a transcendental equation similar to the one just written. The TE modes are treated in a similar way, but in this case the *derivatives* of J_n and Y_n are involved.

4.5.2 TEM Mode for the Coaxial Metallic Cable

Let us study now the TEM mode for a coaxial cable of inner radius a and outer radius b, for which the functional dependence is quite different. In this case the transverse Laplace equation, in polar coordinates needs to be processed, which, separating the variables becomes:

$$\frac{\rho}{P}\frac{d}{d\rho}\left(\rho\frac{dP}{d\rho}\right) + \frac{1}{\phi}\frac{d^2\phi}{d\varphi^2} = 0.$$

The boundary condition is $\frac{\partial T}{\partial s} = 0$ on s, where $ds = a\,d\varphi$ or $ds = b\,d\varphi$, respectively on the inner and outer circumference, so

$$\frac{P(a)}{a}\frac{d\phi}{d\varphi} = 0 \qquad \frac{P(b)}{b}\frac{d\phi}{d\varphi} = 0,$$

The previous equations might be satisfied if $P(a) = P(b) = 0$, but this would mean $T = 0$ on the whole contour, while a TEM mode can propagate only if T assumes (constant) different values on the two contours. This means that either $P(a)$ or $P(b)$ must be non-zero, and then it must be $\frac{d\phi}{d\varphi} \equiv 0$, $\phi(\varphi) = C_2$, and the differential equation becomes:

$$\frac{\rho}{P}\frac{d}{d\rho}\left(\rho\frac{dP}{d\rho}\right) = 0,$$

$$\Rightarrow \rho\frac{dP}{d\rho} = constant = D \Rightarrow \frac{dP}{d\rho} = \frac{D}{\rho},$$

$$\Rightarrow \frac{dT}{d\rho} = C_2\frac{D}{\rho} = \frac{C}{\rho},$$

(so the function $T(\rho)$ is a natural logarithm).

At this point let us recall the expression of the transverse gradient in polar coordinates:

$$\nabla_t = \underline{\rho}_o\frac{\partial}{\partial\rho} + \underline{\varphi}_o\frac{1}{\rho}\frac{\partial}{\partial\varphi},$$

that becomes in our case $\nabla_t = \underline{\rho}_o\frac{d}{d\rho}$, so:

$$\nabla_t T = \underline{\rho}_o\frac{C}{\rho} \quad \propto \quad \underline{e}_t,$$

$$\nabla_t T \times \underline{z}_o = -\frac{C}{\rho}\underline{\varphi}_o \quad \propto \quad \underline{h}_t.$$

In conclusion, the electric field in the cable section is purely radial, and its amplitude is inversely proportional to the distance from the center: it is the same behavior of the electrostatic field in a cylindrical infinite capacitor. On the other hand, the magnetic field is purely circumferential, and its amplitude is again inversely proportional to the distance from the center: it is the same behavior of the magnetostatic field in the region between two cylindrical coaxial conductors in which opposite constant currents flow.

4.6 Cavity Resonators

Let us now briefly discuss the fields that can exist in closed regions, limited by perfectly conductive (metal) walls and filled with a homogeneous, isotropic and non-dispersive, but generically dissipative medium and in the absence of sources. These regions are called cavity resonators (note that there are also open dielectric resonators: there is the same relationship as between closed metal waveguides and open dielectric waveguides). The electromagnetic field solutions, satisfying the boundary conditions, form the free oscillations of the system and they are called resonant or oscillation modes of the cavity.

From a mathematical point of view this is still an eigenvalue problem, but this time three-dimensional, represented by the Helmholtz equation (for example for the electric field) $\nabla^2 \underline{E} + k^2 \underline{E} = 0 \Rightarrow -\nabla^2 \underline{E} = k^2 \underline{E}$. This time, pay attention, $k^2 = \omega^2 \mu \varepsilon_c = \omega^2 \mu \varepsilon - j\omega\mu\sigma$ is considered an eigenvalue to be determined, and not an assigned quantity, as it was in waveguide analysis. Now, in fact, the frequency is the unknown of the problem, while it was k_z for waveguides.

Note that, since the walls are perfectly conductive, the tangential component of the electric field over the whole surface S of the resonator is assigned (and it is zero). However, we have seen that the uniqueness theorem is not valid (and therefore different solutions, besides the trivial one identically zero, could also exist), in two situations:

1. when the medium is non-dissipative and the frequency assumes certain particular values, which are solutions of

$$\int_V (w_H - w_E)\, dV = 0;$$

2. when the medium is dissipative, for certain complex values of ω.

It is possible to demonstrate that $-\nabla^2$ operator is self-adjoint and positive semi-definite (the zero eigenvalue corresponds to static solutions, $\omega = 0$). The eigenvalues are also here a countable infinity (characteristic of closed regions), to which corresponds a countable infinite number of oscillation modes, each with its own resonant frequency, although there may exist different, but degenerate, modes having the same resonance frequency.

In the case of a non-dissipative medium it is simply $\omega = k/\sqrt{\mu\varepsilon}$; in the dissipative case, known k^2, which is always real since the operator is self-adjoint, the complex $\omega = \omega_R + j\omega_j$ can be calculated from $\omega^2 \mu\varepsilon - j\omega\mu\sigma - k^2 = 0$:

$$\Rightarrow \omega^2 - j\frac{\sigma}{\varepsilon}\omega - \frac{k^2}{\mu\varepsilon} = 0,$$

$$\omega = \frac{j\frac{\sigma}{\varepsilon} \pm \sqrt{-\frac{\sigma^2}{\varepsilon^2} + 4\frac{k^2}{\mu\varepsilon}}}{2} = j\frac{\sigma}{2\varepsilon} \pm \sqrt{-\frac{\sigma^2}{4\varepsilon^2} + \frac{k^2}{\mu\varepsilon}}.$$

However, in general the values of practical interest are the ones satisfying:

$$\frac{k^2}{\mu\varepsilon} \gg \frac{\sigma^2}{4\varepsilon^2} \Rightarrow \omega_R = \sqrt{\frac{k^2}{\mu\varepsilon} - \frac{\sigma^2}{4\varepsilon^2}} \rightarrow e^{j\omega_R t},$$

$$w_j = \frac{\sigma}{2\varepsilon} \rightarrow e^{-\omega_j t};$$

in which the last term represents the oscillation damping resulting from the power dissipation.

The figure of merit or quality factor Q is a dimensionless parameter that is usually introduced in the presence of damped oscillations, and defined by:

$$Q = \frac{\omega_R}{2\omega_j} = \frac{\omega_R \varepsilon}{\sigma},$$

it follows that $Q \rightarrow \infty$ for $\sigma \rightarrow 0$. Note that the Q definition just presented is very general, and it is applicable also when losses of different type lead to a complex angular frequency, for example losses due to the finite conductivity of the walls (in this case the boundary conditions on the walls change and in general k^2 is no longer real), or due to openings in the walls (radiation losses, which always exist in the case of open dielectric resonators). In this case the expression of w_j is more complicated and requires the determination of the flux of the Poynting vector through the walls (power dissipated in the walls themselves or radiated through the openings).

The expression of the Poynting's theorem takes the following form for a complex ω and in the frequency domain:

$$\frac{1}{2} \oint_S \underline{n} \cdot (\underline{E} \times \underline{H}^*) \, dS + \frac{1}{2} \int_V \underline{J}_c^* \cdot \underline{E} \, dV + j\frac{1}{2} \int_V (\omega \underline{B} \cdot \underline{H}^* - \omega^* \underline{E} \cdot \underline{D}^*) \, dV =$$
$$= -\frac{1}{2} \int_V (\underline{J}_i^* \cdot \underline{E} + \underline{J}_{mi} \cdot \underline{H}^*) \, dV.$$

In the case of a dissipative homogeneous isotropic and non-dispersive cavity, and in the hypothesis of absence of impressed electric and magnetic currents, taking the real part, it is:

$$\text{Re}\left[\oint_S \frac{1}{2} \underline{n} \cdot (\underline{E} \times \underline{H}^*) \, dS \right] + \int_V \frac{1}{2} \sigma \underline{E} \cdot \underline{E}^* \, dV - 2\omega_j \int_V \frac{1}{4} (\mu \underline{H} \cdot \underline{H}^* + \varepsilon \underline{E} \cdot \underline{E}^*) \, dV = 0,$$

$$\Rightarrow 2\omega_j = \frac{\text{Re}\left[\oint_S \frac{1}{2} \underline{n} \cdot (\underline{E} \times \underline{H}^*) \, dS \right] + \int_V \frac{1}{2} \sigma \underline{E} \cdot \underline{E}^* \, dV}{\int_V \frac{1}{4} (\mu \underline{H} \cdot \underline{H}^* + \varepsilon \underline{E} \cdot \underline{E}^*) \, dV}.$$

In the assumption $\omega_j \ll \omega_R$ the term in square brackets represents the average power flowing in a quasi-period (we are dealing with a damped phenomenon, so not exactly periodic) $T = \frac{2\pi}{\omega_R}$ (for $t = 0$) through the cavity walls (such power is dissipated in the non-perfectly conductive walls or radiated through the openings). The second term at the numerator represents the average power dissipated in the medium (which is not perfect dielectric) that fills the cavity. Therefore, the numerator represents the total average power P lost by the resonator. On the other hand the denominator represents the average energy (magnetic and electric) W stored in the resonator. We therefore have:

$$2\omega_j = \frac{P}{W} \Rightarrow Q = \frac{\omega_R W}{P}.$$

Note that both variables W and P vary over time as $e^{-2\omega_j t}$.

4.6.1 Cylindrical Resonator

We are going to consider now a particularly simple case of resonator, the so-called cylindrical resonator, obtained by closing a stretch of metallic waveguide of length l with two transverse metallic plates (Fig. 4.6).

The simplicity of the problem lies in the fact that one can use the results obtained when studying the waveguide problem. In fact, the oscillation modes of the cylindrical resonator can be obtained from the propagation modes in the guide imposing the additional boundary conditions required by the presence of the two transverse plates, i.e. imposing: $\underline{E}_t = 0$ for $z = 0$ and $z = l$.

Let us consider first TM modes in the guide, for which we had:

$$E_z(q_1, q_2, z) = e_z(q_1, q_2)\, I(z),$$

it follows

$$\underline{E}_t(q_1, q_2, z) = \underline{e}_t(q_1, q_2)\, V(z) \propto \underline{e}_t \frac{dI}{dz}.$$

So $\frac{dI}{dz} = 0$ must be imposed for $z = 0$ and $z = l$, where, for $k_z \neq 0$:

$$I(z) = I_o^+ e^{-jk_z z} + I_o^- e^{jk_z z}$$

while for $k_z = 0$:

$$I(z) = I_{01}\, z + I_{02}.$$

So if $k_z \neq 0$ we have:

$$\frac{dI}{dz} = jk_z\,(-I_o^+ e^{-jk_z z} + I_o^- e^{jk_z z}) = 0$$

Fig. 4.6 Cylindrical resonator

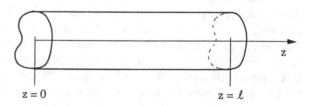

$$z = 0 \qquad\qquad\qquad z = l$$

when $z = 0, l$. Assuming $k_z = 0$ we have instead:

$$\frac{dI}{dz} = I_{01} = 0 \Rightarrow I(z) = I_{02}.$$

In the first case it is obtained, for $z = 0$, $I_o^- = I_o^+$

$$\Rightarrow \frac{dI}{dz} = jk_z I_o^+ 2j \sin(k_z z),$$

and so we have, for $z = l$:

$$-2k_z I_o^+ \sin(k_z l) = 0$$

$$\Rightarrow k_z = \frac{p\pi}{l} \quad p = 1, 2, 3, \ldots,$$

$$\Rightarrow I(z) = 2I_o^+ \cos(\frac{p\pi}{l} z) \quad p = 1, 2, 3, \ldots .$$

Finally the case $k_z = 0$ can be included in the $k_z \neq 0$ one allowing the index p to assume also the zero value.

In the case of TE modes it is:

$$H_z(q_1, q_2, z) = h_z(q_1, q_2) V(z),$$

which yields then:

$$\underline{E}_t(q_1, q_2, z) = \underline{e}_t(q_1, q_2) V(z).$$

We must impose $V(z) = 0$ for $z = 0$, $z = l$, in $V(z) = V_o^+ e^{-jk_z z} + V_o^- e^{jk_z z}$ if $k_z \neq 0$, and in $V(z) = V_{01} z + V_{02}$ if $k_z = 0$. In the latter case it follows $V_{02} = 0$ and $V_{01} = 0$, i.e. the trivial solution is obtained. In the case $k_z \neq 0$ for $z = 0$ it is instead:

$$V_o^- = -V_o^+,$$

$\Rightarrow V(z) = -V_o^+ 2j \sin(k_z z)$, and for $z = l$ we have $\sin(k_z l) = 0 \Rightarrow k_z = \frac{p\pi}{l}$ $p = 1, 2, 3, \ldots$. This time the null value, as said, cannot be taken.

The eigenvalue k^2 is now given by $k^2 = k_t^2 + k_z^2 = k_t^2 + \left(\frac{p\pi}{l}\right)^2$ where the k_t^2 was already found when studying the analogous problem in the guide. For example, for a parallelepiped resonator with a, b and l sides it is:

$$k_{mnp}^2 = \left(\frac{m\pi}{a}\right)^2 + \left(\frac{n\pi}{b}\right)^2 + \left(\frac{p\pi}{l}\right)^2,$$

where, as we have seen, one of the three indices can be zero at most (m or n in TE case, p in TM case).

Chapter 5
Green's Functions

Abstract The solution of Helmholtz equation with a source term is faced, introducing the fundamental concept of Green's function. The basic definitions and algebraic properties are reported, inverse and adjoint operators and eigenfunction expansions are introduced. For the important case of free space, the relevant boundary conditions at infinity are discussed, the Sommerfeld radiation condition described and the scalar Green's functions obtained. Finally, some elements about the presence of metal bodies are given.

5.1 Non-homogeneous Helmholtz Equation, Deterministic Problem

Let us recall the electromagnetic potential concepts exposed previously, in the absence of impressed magnetic currents:

$$\underline{J}_{mi} = 0 \quad \Longrightarrow \quad \nabla^2 \underline{A} + k^2 \underline{A} = -\underline{J}_i \, .$$

We use Cartesian coordinates and project the above equation on the x, y and z axes, obtaining:

$$\nabla^2 A_x + k^2 A_x = -J_{i_x}$$
$$\nabla^2 A_y + k^2 A_y = -J_{i_y}$$
$$\nabla^2 A_z + k^2 A_z = -J_{i_z} \, .$$

Therefore the vector differential equation is converted into three scalar equations, in each of them only one component appears. Note that A_z only could be separated using cylindrical coordinates, and no components could be separated using spherical coordinates. The above separation permitted us to break the vectorial problem into three independent scalar problems.

Such problems present the mathematical form of the so-called *deterministic problem*:

$$Lf = h \, ,$$

© Springer International Publishing Switzerland 2015
F. Frezza, *A Primer on Electromagnetic Fields*,
DOI 10.1007/978-3-319-16574-5_5

being in our case, e.g. for the x component:

$$L = - \left(\nabla^2 + k^2 \right) , \qquad f = A_x , \qquad h = J_{ix} .$$

A_x needs to satisfy boundary conditions which define the domain of the linear operator L. The solution of the deterministic problem is essentially given by the inversion of the L operator:

$$f = L^{-1} h$$

having defined the inverse operator with L^{-1}, it follows $L^{-1} L = L L^{-1} = I$ where I is called identity operator, because, when applied to any function, it returns as a result the function itself.

It can be demonstrated that necessary and sufficient condition for the problem to have a unique solution is that the corresponding homogeneous problem $Lf = 0$ has no eigensolutions (i.e., not identically zero solutions which are eigenfunctions of the zero eigenvalue, and which constitute a vectorial space called *kernel* of the L operator). In fact, assuming the existence of two distinct solutions f_1 and f_2 it follows:

$$Lf_1 = h \ , \quad Lf_2 = h \qquad \Longrightarrow \qquad L(f_1 - f_2) = 0$$

and hence an eigenfunction (not identically zero) associated with the eigenvalue zero exists, i.e. the zero eigenvalue is present. If, conversely, there is an eigenfunction (not identically zero) associated with the zero eigenvalue, i.e., such that $Lf_o = 0$, then for linearity:

$$L(f + f_o) = h$$

is another solution, different from f, of our deterministic problem.

This condition can also be expressed with reference to the adjoint operator, being the eigenvalues of the adjoint operator the complex conjugate of the ones of the L operator. In fact, if we consider the following two eigenvalue problems:

$$L\varphi_i = \lambda_i \, \varphi_i$$
$$L^a \varphi_j^a = \lambda_j^a \, \varphi_j^a ,$$

we have, from the definition of adjoint operator:

$$\left\langle L\varphi_i , \varphi_j^a \right\rangle = \lambda_i \left\langle \varphi_i , \varphi_j^a \right\rangle = \left\langle \varphi_i , L^a \varphi_j^a \right\rangle = \lambda_j^{a*} \left\langle \varphi_i , \varphi_j^a \right\rangle$$

$$\Downarrow$$

$$(\lambda_i - \lambda_j^{a*}) \left\langle \varphi_i , \varphi_j^a \right\rangle = 0 .$$

Now, if it was $\lambda_j^{a*} \neq \lambda_i$ for any choice of i, it would result that the corresponding eigenfunction φ_j^a would be orthogonal to all the φ_i, which form, as we already showed, a complete set. So φ_j^a would be identically zero, against the hypothesis that it is an eigenfunction of L^a. Hence there must be a value of i for which $\lambda_j^{a*} = \lambda_i$. It has also been shown that the eigenfunctions of the operator and those of its adjoint are orthogonal when the corresponding eigenvalues are not conjugated.

Then we can also write:

$$L^a f_o^a = 0$$

to express the condition of uniqueness of the deterministic problem, because, as already pointed out, the operator L has the zero eigenvalue if and only if L^a has that eigenvalue, too.

5.2 Definition and Properties of the Green's Function

Assuming that the deterministic problem has a unique solution, we will solve it using the method of the so-called *Green's function*.

Let us start from the scalar problem, for simplicity, defining the Green's function as follows:

$$L\, G(\underline{r}, \underline{r}') = \delta\left(\underline{r} - \underline{r}'\right),$$

where the expression of the Dirac function δ in Cartesian coordinates is:

$$\delta(\underline{r} - \underline{r}') = \delta(x - x')\, \delta(y - y')\, \delta(z - z')$$

(physical dimensions of the inverse of a volume). The function G must belong to the domain of the operator L, i.e. must satisfy the boundary conditions of the problem. The $\underline{r}$ and $\underline{r}'$ points are in general called *observation point* and *source point*.

G is therefore the spatial impulse response of our electromagnetic *system* (defined by the differential equation and associated boundary conditions), i.e., in our case, the component of vector potential in a certain direction, produced by a spatial pulse of current in the same direction.

Let us define the Green's function G^a for the adjoint operator L^a too, such that:

$$L^a G^a(\underline{r}, \underline{r}') = \delta(\underline{r} - \underline{r}').$$

Note that the Green's function can not exist if the solution of the deterministic problem is not unique. In fact, in case of existence of the Green's function, we would have:

$$\langle L\, G(\underline{r}, \underline{r}'), f_o^a(\underline{r}) \rangle = \langle \delta(\underline{r} - \underline{r}'), f_o^a(\underline{r}) \rangle = f_o^{a*}(\underline{r}') =$$
$$= \langle G(\underline{r}, \underline{r}'), L^a\, f_o^a(\underline{r}) \rangle =$$
$$= \langle G(\underline{r}, \underline{r}'), 0 \rangle \equiv 0$$

against the hypothesis that f_o^a is not identically zero. The two functions $G(\underline{r}, \underline{r}')$ and $G^a(\underline{r}, \underline{r}')$ are not independent because each of them may be derived from the other. In fact, let us consider the relations:

$$L\, G(\underline{r}, \underline{r}_1) = \delta(\underline{r} - \underline{r}_1)$$
$$L^a\, G^a(\underline{r}, \underline{r}_2) = \delta(\underline{r} - \underline{r}_2) \, ,$$

where $\underline{r}_1$ and $\underline{r}_2$ are source points of L and L^a respectively. Let us demonstrate that $G(\underline{r}, \underline{r}') = G^{a*}(\underline{r}', \underline{r})$:

$$\langle L\, G(\underline{r}, \underline{r}_1), G^a(\underline{r}, \underline{r}_2)\rangle = \langle \delta(\underline{r} - \underline{r}_1), G^a(\underline{r}, \underline{r}_2)\rangle = G^{a*}(\underline{r}_1, \underline{r}_2) =$$
$$= \langle G(\underline{r}, \underline{r}_1), L^a\, G^a(\underline{r}, \underline{r}_2)\rangle =$$
$$= \langle G(\underline{r}, \underline{r}_1), \delta(\underline{r} - \underline{r}_2)\rangle =$$
$$= G(\underline{r}_2, \underline{r}_1)$$

being the function δ real.

One way to compute the Green's function is to integrate directly the differential equation, this method will be shown later. Another way to deal with the problem is to represent G by a series expansion of the eigenfunction set

$$G(\underline{r}, \underline{r}') = \sum_{n=1}^{\infty} c_n(\underline{r}')\, \varphi_n(\underline{r}) \, ,$$

where the c_n coefficients are generally functions of $\underline{r}'$. It follows:

$$L\, G(\underline{r}, \underline{r}') = L \sum_{n=1}^{\infty} c_n(\underline{r}')\, \varphi_n(\underline{r}) = \sum_{n=1}^{\infty} c_n(\underline{r}')\, L\, \varphi_n(\underline{r}) =$$
$$= \sum_{n=1}^{\infty} c_n(\underline{r}')\, \lambda_n\, \varphi_n(\underline{r}) = \delta(\underline{r} - \underline{r}')$$

assuming the possibility of permuting the operator L with the series, i.e. of deriving the series term by term.

Now let's consider the following scalar product in which φ_m^a is an eigenfunction of L^a, corresponding to the $\lambda_m^a = \lambda_m^*$ eigenvalue and normalized with respect to φ_m (i.e., such that $\langle \varphi_m, \varphi_m^a\rangle = 1$):

$$\langle L\, G(\underline{r}, \underline{r}'), \varphi_m^a(\underline{r})\rangle = \left\langle \sum_n c_n(\underline{r}')\, \lambda_n\, \varphi_n(\underline{r}), \varphi_m^a(\underline{r})\right\rangle$$
$$= \sum_n c_n(\underline{r}')\, \lambda_n\, \langle \varphi_n(\underline{r}), \varphi_m^a(\underline{r})\rangle =$$
$$= c_m(\underline{r}')\, \lambda_m\, \langle \varphi_m(\underline{r}), \varphi_m^a(\underline{r})\rangle = c_m(\underline{r}')\, \lambda_m =$$

$$= \langle \delta(\underline{r} - \underline{r}'), \varphi_m^a(\underline{r}) \rangle =$$
$$= \varphi_m^{a*}(\underline{r}'),$$

so:

$$c_m(\underline{r}') = \frac{\varphi_m^{a*}(\underline{r}')}{\lambda_m} \qquad \Longrightarrow \qquad G(\underline{r}, \underline{r}') = \sum_{n=1}^{\infty} \frac{\varphi_n(\underline{r}) \, \varphi_n^{a*}(\underline{r}')}{\lambda_n}.$$

Moreover, the δ function can be also expressed as:

$$\delta(\underline{r} - \underline{r}') = \sum_{n=1}^{\infty} \varphi_n(\underline{r}) \, \varphi_n^{a*}(\underline{r}').$$

The two previous series expansions are called *spectral representations* of the Green's function and of the δ function respectively.

From the knowledge of the Green function of the adjoint operator, it is possible to compute the inverse of the operator L and then to solve the deterministic problem. Let us consider the following scalar product:

$$\langle f(\underline{r}), L^a \, G^a(\underline{r}, \underline{r}') \rangle = \langle f(\underline{r}), \delta(\underline{r} - \underline{r}') \rangle = f(\underline{r}') =$$
$$= \langle L \, f(\underline{r}), G^a(\underline{r}, \underline{r}') \rangle = \langle h(\underline{r}), G^a(\underline{r}, \underline{r}') \rangle =$$
$$= \int_V h(\underline{r}) \, G^{a*}(\underline{r}, \underline{r}') \, dV = \int_V h(\underline{r}) \, G(\underline{r}', \underline{r}) \, dV,$$

and exchanging the role of the variables with and without prime

$$f(\underline{r}) = \int_V G(\underline{r}, \underline{r}') \, h(\underline{r}') \, dV'.$$

This resolutive solution could also be found by applying the superposition principle, starting from

$$L \, G(\underline{r}, \underline{r}') = \delta(\underline{r} - \underline{r}'),$$

multiplying by $h(\underline{r}')$ and then integrating on V with respect to the variable $\underline{r}'$:

$$\int_V h(\underline{r}') \, L \, G(\underline{r}, \underline{r}') \, dV' = \int_V h(\underline{r}') \, \delta(\underline{r} - \underline{r}') \, dV'.$$

Now L can be taken out of the integral, because it does not operate on $\underline{r}'$:

$$L \int_V G(\underline{r}, \underline{r}') \, h(\underline{r}') \, dV' = h(\underline{r}),$$

but being $L \, f = h$, it is:

$$f(\underline{r}) = \int_V G(\underline{r}, \underline{r}') \, h(\underline{r}') \, dV'.$$

Hence:

$$L^{-1}[\,] = \int_V G(\underline{r}, \underline{r}')[\,] \, dV'.$$

In the case in which the problem is inherently vectorial (because the boundary conditions can not be separated for the various components), we need to introduce a dyadic Green's function: $\underline{\underline{G}}(\underline{r}, \underline{r}')$ such that:

$$L \, \underline{\underline{G}}(\underline{r}, \underline{r}') = \underline{\underline{I}} \, \delta(\underline{r} - \underline{r}'),$$

the unit dyad being defined by $\underline{\underline{I}} = \left(\begin{smallmatrix} 1 & 0 & 0 \\ 0 & 1 & 0 \\ 0 & 0 & 1 \end{smallmatrix}\right)$, so that, when applied to a vector $\underline{A}$ or a dyad $\underline{\underline{D}}$ it is:

$$\underline{\underline{I}} \cdot \underline{A} = \underline{A} \cdot \underline{\underline{I}} = \underline{A} \quad \text{and} \quad \underline{\underline{I}} \cdot \underline{\underline{D}} = \underline{\underline{D}} \cdot \underline{\underline{I}} = \underline{\underline{D}}$$

respectively. Moreover, the unit dyad in Cartesian coordinates assumes the form

$$\underline{\underline{I}} = \underline{x}_o \underline{x}_o + \underline{y}_o \underline{y}_o + \underline{z}_o \underline{z}_o.$$

Proceeding in a manner conceptually similar to the scalar case, the resolutive solution of the vectorial deterministic problem can be obtained:

$$\underline{f}(\underline{r}) = \int_V \underline{\underline{G}}(\underline{r}, \underline{r}') \cdot \underline{h}(\underline{r}') \, dV'.$$

5.3 Boundary Conditions at Infinity for Free Space

We now want to determine the vector potential produced by a space-limited distribution of currents in free space occupied by a linear, stationary, homogeneous and isotropic medium.

We need to assume boundary conditions in free space too, called conditions at infinity. So, said r the radius in spherical coordinates, it must be, e.g. for the A_x component of the potential vector:

$$\lim_{r \to \infty} (r|A_x|) = \ell < \infty,$$

and the same for A_y and A_z components. This means that, for $r \to \infty$, $|A_x|$ goes to zero as fast as $\frac{1}{r}$ at minimum. This follows on the hypothesis that the impressed currents do not extend to infinity.

Let us impose now an additional constraint, said *Sommerfeld condition* or radiation condition, assuming that the field is analogous to a spherical wave expanding in space, i.e. we assume a radial dependence $\frac{e^{-jkr}}{r}$ (field energy keeps finite), hence the wave is propagating radially away from the sources which are space-limited. It is then:

$$\lim_{r \to \infty} \left[r \left(\frac{\partial A_x}{\partial r} + jk\, A_x \right) \right] = 0$$

and similar relations for A_y and A_z. The Sommerfeld condition is more restrictive than the former one: in fact, it can be shown that any solution of the differential equation that satisfies the Sommerfeld condition also satisfies the other one, and therefore it is sufficient to impose the Sommerfeld condition.

Since the boundary conditions seen are separated for the various components, the deterministic problem assumes a scalar form. The domain of the operator (from which some of the operator properties depend) is determined by the boundary conditions. In the examined case the operator

$$L = -(\nabla^2 + k^2)$$

is not self-adjoint. This is due essentially to the presence of the constant k^2, which is in general complex, and to the factor j in the Sommerfeld condition. It can be demonstrated the existence of an adjoint operator L^a, expressed by:

$$L^a = -(\nabla^2 + k^{2*})$$

and its domain is defined by the g functions such that:

$$\lim_{r \to \infty} (r\, |g|) = \ell < \infty$$

and

$$\lim_{r \to \infty} \left[r \left(\frac{\partial g}{\partial r} - jk^* g \right) \right] = 0.$$

In fact it is:

$$\langle L\, f, g \rangle = - \int_V (\nabla^2 f + k^2\, f)\, g^* \, dV =$$

$$= \int_V \nabla f \cdot \nabla g^* \, dV - \oint_S \frac{\partial f}{\partial n} g^* \, dS - k^2 \int_V f\, g^* \, dV \,,$$

having applied the first form of Green's lemma. On the other hand, it is:

$$\langle f, L^a\, g \rangle = - \int_V f \left(\nabla^2 g + k^{2*}\, g \right)^* \, dV =$$

$$= -\int_V f \left(\nabla^2 g^* + k^2 g^* \right) dV =$$

$$= \int_V \nabla f \cdot \nabla g^* \, dV - \oint_S f \frac{\partial g^*}{\partial n} \, dS - k^2 \int_V f g^* \, dV .$$

Therefore, it is finally:

$$\langle L f, g \rangle - \langle f, L^a g \rangle = \oint_S \left(f \frac{\partial g^*}{\partial n} - \frac{\partial f}{\partial n} g^* \right) dS .$$

Let us perform now, for simplicity, the surface integral on a sphere having center at the origin and radius going to infinity, the area element is in spherical coordinates:

$$dS = r^2 \sin \theta \, d\theta \, d\varphi .$$

Moreover, the normal derivative coincides with the radial derivative:

$$\int_0^{2\pi} \int_0^\pi \left(f \frac{\partial g^*}{\partial r} - \frac{\partial f}{\partial r} g^* \right) r^2 \sin \theta \, d\theta \, d\varphi .$$

Let us observe now that from the condition for the adjoint operator it follows, conjugating:

$$\lim_{r \to \infty} \left[r \left(\frac{\partial g^*}{\partial r} + jk \, g^* \right) \right] = 0$$

$$\Downarrow$$

$$\lim_{r \to \infty} \left(\frac{\partial g^*}{\partial r} + jk \, g^* \right) = 0$$

$$\Downarrow$$

$$\lim_{r \to \infty} \frac{\partial g^*}{\partial r} = \lim_{r \to \infty} -jk \, g^* ,$$

and multiplying by $r^2 f$:

$$\lim_{r \to \infty} \left(r^2 f \frac{\partial g^*}{\partial r} \right) = \lim_{r \to \infty} (-jk \, r^2 \, f \, g^*) .$$

Moreover, from the boundary condition on the operator L, written for the function f, it follows in a similar manner, multiplying by $r^2 g^*$:

$$\lim_{r \to \infty} \left(r^2 \frac{\partial f}{\partial r} g^* \right) = \lim_{r \to \infty} (-jk \, r^2 \, f \, g^*)$$

and since such limits are finite for the condition on the magnitudes of f, g (so that it is avoided the occurrence of $\infty - \infty$ indeterminate form), the integrand is zero for $r \to \infty$ and therefore L^a is the adjoint of L as assumed.

The fact that the expression for the adjoint operator and its condition at infinity are obtained from the corresponding ones for the operator L, simply by replacing the complex constants with their conjugates, is of general validity (it is worth recalling that ∇^2 operator is real, and therefore it returns real functions when it is applied to real functions).

5.4 Green's Function Calculation for the Helmholtz Equation in Free Space

The differential equation to be solved is:

$$-(\nabla^2 + k^2)\, G(\underline{r}, \underline{r}') = \delta(\underline{r} - \underline{r}')$$

with the boundary condition at infinity.

Let us choose the source point $\underline{r}'$ (in which the impulsive source is centered) in the origin of our reference system for simplicity and symmetry (and, moreover, without loss of generality), then we have:

$$-(\nabla^2 + k^2)\, G(\underline{r}) = \delta(\underline{r})\,.$$

Assuming now a spherical reference system r, θ, φ, with the origin in the above source point $\underline{r}'$, we have, for the spherical symmetry of both free space and the source $\delta(\underline{r}) = \delta(r)$, that Green's function depends only on r, and therefore we have:

$$\frac{\partial G}{\partial \theta} = \frac{\partial G}{\partial \varphi} = 0 \quad \Longrightarrow \quad G(\underline{r}) = G(r)\,.$$

The Green's function is therefore a uniform spherical wave, because both the equiphase and the equiamplitude surfaces are represented by spheres ($r = constant$).

Let us consider now the expression of ∇^2 operator in spherical coordinates:

$$\nabla^2 G = \frac{1}{r^2}\frac{\partial}{\partial r}\left(r^2 \frac{\partial G}{\partial r}\right) + \frac{1}{r^2 \sin\theta}\frac{\partial}{\partial \theta}\left(\sin\theta \frac{\partial G}{\partial \theta}\right) + \frac{1}{r^2 \sin^2\theta}\frac{\partial^2 G}{\partial \varphi^2}\,,$$

where the second and third terms cancel out for the assumptions made above, so:

$$-\left[\frac{1}{r^2}\frac{d}{dr}\left(r^2 \frac{dG}{dr}\right) + k^2\, G\right] = \delta(r)\,.$$

We are initially searching for a solution for $r \neq 0 \Rightarrow \delta(r) = 0$. Multiplying the equation by r, we obtain:

$$\frac{1}{r} \frac{d}{dr} \left(r^2 \frac{dG}{dr} \right) + k^2 r G = 0.$$

Let's consider now the auxiliary function: $\widetilde{G}(r) = r\, G(r)$

$$\Rightarrow \frac{d\widetilde{G}}{dr} = G + r \frac{dG}{dr} \qquad \Longrightarrow \qquad r \frac{dG}{dr} = \frac{d\widetilde{G}}{dr} - G$$

and, multiplying by r:

$$r^2 \frac{dG}{dr} = r \frac{d\widetilde{G}}{dr} - r\, G = r \frac{d\widetilde{G}}{dr} - \widetilde{G}.$$

Recalling the starting equation:

$$\frac{1}{r} \frac{d}{dr} \left(r \frac{d\widetilde{G}}{dr} - \widetilde{G} \right) + k^2 \widetilde{G} = 0$$

and performing the derivative we obtain:

$$\frac{1}{r} \left(\frac{d\widetilde{G}}{dr} + r \frac{d^2\widetilde{G}}{dr^2} - \frac{d\widetilde{G}}{dr} \right) + k^2 \widetilde{G} = 0$$

$$\Downarrow$$

$$\frac{d^2\widetilde{G}}{dr^2} + k^2 \widetilde{G} = 0.$$

It is apparent that we have once again obtained the one-dimensional Helmholtz equation. Therefore, when $k \neq 0$:

$$\widetilde{G}(r) = C_1 e^{-jkr} + C_2 e^{jkr} \qquad \Longrightarrow \qquad G(r) = C_1 \frac{e^{-jkr}}{r} + C_2 \frac{e^{jkr}}{r}.$$

The Sommerfeld boundary condition is imposed now, so:

$$r \frac{dG}{dr} + jkr\, G = \frac{d\widetilde{G}}{dr} - G + jk\, \widetilde{G} =$$

$$= jk \left(-C_1 e^{-jkr} + C_2 e^{jkr} \right) - \left(C_1 \frac{e^{-jkr}}{r} + C_2 \frac{e^{jkr}}{r} \right) +$$

$$+ jk \left(C_1 e^{-jkr} + C_2 e^{jkr} \right) =$$

$$= -C_1 \frac{e^{-jkr}}{r} + C_2 e^{jkr} \left(2jk - \frac{1}{r} \right).$$

Assuming $k = \beta - j\alpha$ it will be $\alpha \geq 0$ for passive media and then the above quantity will go to zero when $r \to \infty$, satisfying the condition, only if $C_2 = 0$

$$\Rightarrow \quad G(r) = C_1 \frac{e^{-jkr}}{r}.$$

The source term needs now to be examined to compute the C_1 constant, still undetermined in the above formula. Let us come back to the:

$$-(\nabla^2 + k^2)\, G = \delta(\underline{r})$$

and let us integrate on the spheric V volume centered in the origin:

$$-\int_V \nabla^2 G\, dV - k^2 \int_V G\, dV = 1.$$

The divergence theorem is then applied, and recalling that $\nabla^2 G = \nabla \cdot \nabla G$:

$$-\oint_S \underline{n} \cdot \nabla G\, dS - k^2 \int_V G\, dV = 1$$

$$\Downarrow$$

$$-\oint_S \frac{\partial G}{\partial n}\, dS - k^2 \int_V G\, dV = 1.$$

Assuming that r_o is the sphere radius:

$$-\frac{dG}{dr}\bigg|_{r=r_o} \oint_S dS - k^2 \int_0^{2\pi} \int_0^{\pi} \int_0^{r_o} G\, r^2 \sin\theta\, dr\, d\theta\, d\varphi = 1$$

having expressed the volume element dV in spherical coordinates.

$$\Rightarrow \quad -4\pi r_o^2 \frac{dG}{dr}\bigg|_{r=r_o} - k^2 C_1 \int_0^{2\pi} d\varphi \int_0^{\pi} \sin\theta\, d\theta \int_0^{r_o} r\, e^{-jkr}\, dr = 1.$$

Now let us perform the derivative and the first two integrals, obtaining:

$$-4\pi r_o^2 C_1 \frac{-jkr_o e^{-jkr_o} - e^{-jkr_o}}{r_o^2} - 4\pi k^2 C_1 \int_0^{r_o} r\, e^{-jkr}\, dr = 1.$$

The last integral can be solved by parts, obtaining:

$$\int_0^{r_o} r\, e^{-jkr}\, dr = \int_0^{r_o} \frac{r(-jk)}{-jk}\, e^{-jkr}\, dr =$$

$$= \frac{r\, e^{-jkr}}{-jk}\bigg|_0^{r_o} - \int_0^{r_o} \frac{1}{-jk}\, e^{-jkr}\, dr =$$

$$= \frac{r_o\, e^{-jkr_o}}{-jk} - \int_0^{r_o} \frac{-jk}{(-jk)^2}\, e^{-jkr}\, dr =$$

$$= \frac{r_o\, e^{-jkr_o}}{-jk} + \frac{1}{k^2}\, e^{-jkr}\bigg|_0^{r_o} =$$

$$= \frac{r_o\, e^{-jkr_o}}{-jk} + \frac{1}{k^2}\left(e^{-jkr_o} - 1\right) =$$

$$= \frac{jkr_o\, e^{-jkr_o} + \left(e^{-jkr_o} - 1\right)}{k^2}$$

$$\Rightarrow \quad 4\pi\, C_1\, e^{-jkr_o}\, (jkr_o + 1) - 4\pi\, C_1 \left[e^{-jkr_o}\, (jkr_o + 1) - 1 \right] = 1$$

$$\Rightarrow \quad 4\pi\, C_1 = 1 \quad \Rightarrow \quad C_1 = \frac{1}{4\pi} \quad \Longrightarrow \quad G(r) = \frac{e^{-jkr}}{4\pi r}.$$

If the source point does not coincide with the origin, i.e. $\underline{r}' \neq 0$, r must be replaced with $|\underline{r} - \underline{r}'|$, i.e. with the distance between the observation point and the source point. In fact, the Green's function in free space is a function of $\underline{r} - \underline{r}'$ and it is not a function of $\underline{r}$ and $\underline{r}'$ taken separate, because free space is a homogeneous structure and so it is invariant to translations. We therefore have:

$$G(\underline{r}, \underline{r}') = \frac{e^{-jk|\underline{r}-\underline{r}'|}}{4\pi|\underline{r} - \underline{r}'|}.$$

Note the important property $G(\underline{r}, \underline{r}') = G(\underline{r}', \underline{r})$.

At this point we can write, for example for the A_x component:

$$A_x(\underline{r}) = \int_V \frac{e^{-jk|\underline{r}-\underline{r}'|}}{4\pi|\underline{r} - \underline{r}'|}\, J_{xi}(\underline{r}')\, dV';$$

similar formulas, as usual, can be written for $A_y(\underline{r})$ and $A_z(\underline{r})$. Multiplying each component by the corresponding Cartesian unit vector, which can be led into the integral as a constant vector, we have the vectorial relation:

$$\underline{A}(\underline{r}) = \int_V \frac{e^{-jk|\underline{r}-\underline{r}'|}}{4\pi|\underline{r} - \underline{r}'|}\, \underline{J}_i(\underline{r}')\, dV'.$$

The magnetic field $\underline{H} = \nabla \times \underline{A}$ can then be derived from this expression obtaining:

$$\underline{H}(\underline{r}) = \nabla \times \int_V G(\underline{r}, \underline{r}') \, \underline{J}_i(\underline{r}') \, dV' =$$

$$= \int_V \nabla \times [G(\underline{r}, \underline{r}') \, \underline{J}_i(\underline{r}')] \, dV' .$$

Now we apply the vector identity

$$\nabla \times (f \, \underline{v}) = f \, \nabla \times \underline{v} - \underline{v} \times \nabla f ,$$

where in our case the vector does not depend on $\underline{r}$, so the integrand becomes

$$\nabla G(\underline{r}, \underline{r}') \times \underline{J}_i(\underline{r}') ;$$

this result could also be quickly obtained just moving the Green's function (scalar) to the other side of the vector product.

The electric field can be computed using the formula:

$$\underline{E} = -j\omega\mu \, \underline{A} + \frac{\nabla \nabla \cdot \underline{A}}{j\omega\varepsilon_c} .$$

The gradient of the divergence can be taken into the integral also in this case, then the following vector identity can be used:

$$\nabla \cdot (f\underline{v}) = \underline{v} \cdot \nabla f + f \, \nabla \cdot \underline{v} .$$

The term $\underline{v} \cdot \nabla f$, i.e. $\underline{J}_i \cdot \nabla G$ is not zero in this case, and its gradient needs to be calculated. To this aim the following identity is exploited:

$$\nabla(\underline{A} \cdot \underline{B}) = \underline{B} \times (\nabla \times \underline{A}) + (\underline{B} \cdot \nabla)\underline{A} + \underline{A} \times (\nabla \times \underline{B}) + (\underline{A} \cdot \nabla)\underline{B} .$$

In our case, the first vector does not depend on $\underline{r}$, and then only the third and fourth term of the identity remain. In particular the nabla operates on the second vector in both terms. Therefore the following integrand is obtained in the second addend providing the electric field:

$$\underline{J}_i(\underline{r}') \times (\nabla \times \nabla G) + \left[\underline{J}_i(\underline{r}') \cdot \nabla \right] \nabla G = \left[\underline{J}_i(\underline{r}') \cdot \nabla \right] \nabla G(\underline{r}, \underline{r}') .$$

It is apparent from the above formulas for the evaluation of the electromagnetic field the importance of the computation of the gradient, and of the gradient of the gradient, of the Green's function.

5.5 Electromagnetic Field Produced by a Distribution of Impressed Currents in the Presence of a Metallic Body

Any perfectly conducting body present in a space region adds a further boundary condition on its surface S: $\underline{n} \times \underline{E} = 0$.

This condition can not be generally decomposed into separate conditions for the $\underline{E}$ components when the shape of S is arbitrary, so the problem must be assumed inherently vectorial. It is convenient in this case the direct usage of the electric field, without passing through a vector potential, as this choice permits to impose more easily the boundary condition on the perfect conductor.

Let us take the curl of the first Maxwell's equation (assuming μ constant):

$$\nabla \times \nabla \times \underline{E} = -\nabla \times \underline{J}_{mi} - j\omega\mu\nabla \times \underline{H} \, ;$$

the second Maxwell's equation can be substituted in the second term on the right, obtaining:

$$\nabla \times \nabla \times \underline{E} = -\nabla \times \underline{J}_{mi} - j\omega\mu \, (\underline{J}_i + j\omega\epsilon_c \, \underline{E}) \, ,$$

from which:

$$\nabla \times \nabla \times \underline{E} - k^2 \underline{E} = -\nabla \times \underline{J}_{mi} - j\omega\mu \, \underline{J}_i \, .$$

This is a form of the wave equation in the frequency domain more general than Helmholtz equation, and the operator is $L = \nabla \times \nabla \times (\cdot) - k^2(\cdot)$ with the boundary conditions on the conducting body and at infinity.

The condition at infinity for the amplitude of the field can be written in this case (assuming that the impressed currents do not extend to infinity):

$$\lim_{r \to \infty} \left(r \, |\underline{E}| \right) = \ell < \infty \, .$$

The radiation condition practically assumes that the field is analogous to a radial plane wave at a great distance from the source, it follows that it is $\underline{E} = \zeta \, \underline{H} \times \underline{r}_o$. It is assumed therefore:

$$\lim_{r \to \infty} \left[r \, (\zeta \, \underline{H} \times \underline{r}_o - \underline{E}) \right] = 0 \, .$$

The above condition can be rewritten using homogeneous Maxwell's equations (because it is assumed that there are no currents in the far region), so that only the electric field $\underline{E}$ appears. Hence, recalling that $\omega\mu = k \, \zeta$:

$$\underline{H} = -\frac{\nabla \times \underline{E}}{j\omega\mu}$$

$$\Rightarrow \lim_{r \to \infty} \left\{ r \left[\frac{\underline{r}_o \times (\nabla \times \underline{E})}{jk} - \underline{E} \right] \right\} = 0$$

$$\Rightarrow \lim_{r \to \infty} \left\{ r \left[\underline{r}_o \times (\nabla \times \underline{E}) - jk\underline{E} \right] \right\} = 0 \, .$$

Scalar multiplying by $\underline{r}_o$ we get for the radial component E_r:

$$\lim_{r \to \infty} (r\, E_r) = 0 \qquad \text{the wave is TE in the radial direction.}$$

It can be verified that in this vector case, too, the Sommerfeld radiation condition implies the condition on the magnitudes, so it is sufficient to impose just the former.

The operator L defined above, with the boundary conditions seen, admits the adjoint operator $L^a = \nabla \times \nabla \times (\cdot) - k^{2*}(\cdot)$ with boundary conditions $\underline{n} \times \underline{g} = 0$ on the conductor surface and

$$\lim_{r \to \infty} \left\{ r \left[\underline{r}_o \times (\nabla \times \underline{g}) + jk^* \underline{g} \right] \right\} = 0 .$$

The problem can be solved through a dyadic Green's function, which has as columns three vector Green's functions which satisfy the boundary conditions of the L operator.

Bibliography

1. G. Gerosa, P. Lampariello, *Lezioni di Campi Elettromagnetici*, 2nd edn. (Edizioni Ingegneria 2000, Roma, 2006)
2. G. Franceschetti, *Electromagnetics: Theory, Techniques, and Engineering Paradigms*, 2nd edn. (Springer, Berlin, 2013)
3. C.G. Someda, *Electromagnetic Waves*, 2nd edn. (CRC, Boca Raton, 2006)
4. J.D. Jackson, *Classical Electrodynamics*, 3rd edn. (Wiley, New York, 1999)
5. C.A. Balanis, *Advanced Engineering Electromagnetics*, 2nd edn. (Wiley, New York, 2012)
6. D.S. Jones, *Acoustic and Electromagnetic Waves* (Oxford University Press, Oxford, 1989)
7. S. Ramo, J.R. Whinnery, T. Van Duzer, *Fields and Waves in Communication Electronics*, 3rd edn. (Wiley, New York, 1994)
8. L.D. Landau, E.M. Lifsits, *Electrodynamics of Continuous Media*, 2nd edn. (Butterworth-Heinemann, Oxford, 1984)
9. G. Barzilai, *Fondamenti di elettromagnetismo* (Siderea, Roma, 1975)
10. A.W. Snyder, J.D. Love, *Optical Waveguide Theory* (Chapman and Hall, London, 1983)
11. I. Cattaneo Gasparini, *Strutture algebriche lineari* (Masson, Milano, 1998)
12. G.C. Corazza, C.G. Someda, *Elementi di calcolo vettoriale e tensoriale* (Pitagora, Bologna, 1982)
13. K. Kurokawa, *An Introduction to the Theory of Microwave Circuits* (Academic Press, New York, 1969)
14. R.F. Harrington, *Time-Harmonic Electromagnetic Fields* (McGraw-Hill, New York, 1961)
15. P.M. Morse, H. Feshbach, *Methods of Theoretical Physics* (McGraw-Hill, New York, 1953)
16. C.T. Tai, *Generalized Vector and Dyadic Analysis: Applied Mathematics in Field Theory* (IEEE Press, New York, 1996)

© Springer International Publishing Switzerland 2015
F. Frezza, *A Primer on Electromagnetic Fields*,
DOI 10.1007/978-3-319-16574-5

Index

A

Admittance
  along the line, 104
  characteristic, 102
  shunt (per unit length), 98
Ampère-Maxwell circulation law, 5

B

Beat velocity, 67–69
Bessel
  equation, 142
  functions, 142, 143
Boundary conditions, 1, 12–14, 16, 24, 35,
    37, 38, 73–76, 78, 82, 94, 97, 103,
    115, 116, 125, 126, 129–131, 134–
    137, 140, 141, 143, 144, 146–148,
    151–153, 156, 157, 159, 161, 164,
    165
Brewster angle, 76, 87, 88, 90, 92
Brillouin
  diagrams, 67
  precursors, 71

C

Cavity resonator, 115, 145–148
  cylindrical, 148
  dielectric, 147
  modes
    TE, 149
    TM, 148, 149
Coaxial cable, 115, 140, 142–145
  TEM mode, 144, 145
Continuity
  conditions, 4, 7, 12, 13, 83, 86, 89, 93
  equations, 5, 6, 8

D

Curl, 1
Cut-off
  condition, 132, 139
  frequency, 132
  wavelength, 132, 139

D'Alembert
  equation, 40, 46
  operator, 40
Dispersion
  diagrams, 67
  equation, 144
  function, 71
  law, 68
  spatial, 11
  temporal, 10
Divergence, 1
Duality, 16, 26, 33, 46, 57, 61, 116, 117
  principle, 7, 14, 41, 42, 45

E

Eigenfunctions, 126–128, 130–132, 136,
    137, 139, 151–154
Eigenvalues, 26, 27, 31, 32, 34, 115, 126–
    132, 134, 136, 138, 144, 146, 152,
    153
Electric
  charge density, 4
  current density, 4
  field, 4
  induction, 4
Electromagnetic potentials, 1, 42–46, 151

© Springer International Publishing Switzerland 2015
F. Frezza, *A Primer on Electromagnetic Fields*,
DOI 10.1007/978-3-319-16574-5

Printed in the United States
By Bookmasters